D1035241

The QUALITY SYSTEM

Frank Caplan

The QUALITY SYSTEM

Second Edition

A Sourcebook for Managers and Engineers

CHILTON BOOK COMPANY
Radnor, Pennsylvania

Published in Radnor, Pennsylvania 19089, by Chilton Book Company

Manufactured in the United States of America

Library of Congress Cataloging in Publication Data

Caplan, Frank, 1919-
The quality system : a sourcebook for managers and engineers /
Frank Caplan.—2nd ed.
p. cm.
ISBN 0-8019-7975-7 (*Chilton*)
ISBN 0-8019-8113-1 (*ASQC*)
1. Quality control. 2. Quality assurance. I. Title.
TS156.C346 1989 89-42860
658.5'62—dc20 CIP

1 2 3 4 5 6 7 8 9 0   8 7 6 5 4 3 2 1 0 9

TO SHIRLEY

without whose constant support and wifely encouragement this book would never have been written.

Contents

Foreword

The Quality System: A Sourcebook for Managers and Engineers meets a critical need in these times of increasing competition for the customer's dollar, lagging productivity, and, perhaps of greater concern, increasing threats of product liability suits. Each manufacturer and supplier of services recognizes the need for adequate systems to assure product quality, to protect the customer from injury, and to protect his business from product liability suits or involvement, whether publicized or not, with the Consumer Product Safety Commission or other agencies.

The quality technology necessary to maximize public satisfaction, to ensure compliance with regulatory requirements, and to assure the profitability and survival of a business enterprise is clearly presented in this excellent book by Frank Caplan. He offers a logically structured package to improve quality, yield, and customer satisfaction, whether the delivered item is a product or service, in the commercial or the government sector. To establish the quality system his business requires, the reader—the quality manager or engineer, the entrepreneur, the quality-minded member of top management—can select whatever elements of the package he wishes. The elements are, in effect and intent, the building blocks of a complete quality system. By studying the 109 quality elements presented, recognizing which ones already exist in his enterprise and assessing their effectiveness, the reader can determine the elements needed to bring his system into line with sound practices of quality management.

This book reflects the author's extensive experience and success in designing and installing effective quality systems in a wide variety of industries and businesses, as a member of line and staff management and as an outside consultant. Mellowed by experience in program and project management both in the United States and abroad, Caplan has put together a theoretically cohesive yet practical approach to a sound quality system. It is based on ideal concepts, but it is modified by practical knowledge of the real world of quality. This volume contains the body and substance of the quality systems I have seen work successfully. The greatest emphasis is on planning, prevention, and control of quality through the careful design and implementation of integrated quality systems. Furthermore, Caplan's system has demonstrated compatibility with the latest and most progressive thinking in quality technology, as recognized in government documents and in the quality standards, ANSI/ASQC A3-1978, *Quality Systems Terminology*, and ANSI/ZI.15-1979, *Generic Guidelines for Quality Systems*.

Throughout, Caplan stresses the *system* aspect of quality control and the essential interactions with other systems and organizations that make quality control a smooth-running, profitable, customer-satisfying operation. His "building block" concept is unique. It offers flexibility and interlocking capability for all necessary elements, regardless of the number and type of elements used in structuring the quality system. The author recognizes that a single quality structure will not work in different organizations and product lines, and that the system must be tailored to the unique requirements of the user. He also acknowledges the aging aspects of systems, encourages the periodic re-examination of systems with product maturity and changing markets, and recommends elimination of tasks no longer needed for the assurance of product quality.

The nine quality subsystems cover the entire life cycle of the product, from product design and development through test, manufacture, marketing, delivery, customer use, and field service. Subsystem 3, Purchased Material Control, is outstanding. Purchases from suppliers and subcontractors account for over 50 percent of the money expended in the average manufacturing enterprise and for even more in many businesses and government agencies. It is one area that requires much more attention than it has been given in the past. This section, like all the others, anticipates all contingencies and offers excellent counsel in problem prevention as well as in problem solving. Caplan's supplier rating system is comprehensive and readily adaptable to most product lines.

Section 10, Quality System Engineering, which follows the nine subsystems, guides the reader through the quality system analysis and design process in easily understandable terms. A format is suggested for use in designing each unique quality system, whether for a product line or for a service. The section on Managing the Quality System shows how to evaluate the overall management and effectiveness of the quality system once it is implemented. The author has provided forms that simplify this task for the quality manager.

This book offers the mature quality manager an excellent tool for assessing the adequacy, scope, and effectiveness of his system. To the moderately experienced manager, it offers an abundance of thought- and action-provoking ideas that he can adopt or adapt to his use. To the fledgling manager, the entrepreneur, the manager of services, it offers a wealth of solid components that can be put together into a quality system that will improve quality control and customer satisfaction, with accompanying improved profit and benefit to the company.

The comprehensiveness, practicality, and proven success of Caplan's system justify the acquisition of this book by every quality engineer and progressive member of management.

Rocco L. Fiaschetti

Preface to the Second Edition

During the ten years since the preparation of the first edition, there have been many changes in the Quality situation in the world. Virtually every newspaper or magazine article, radio and television program or commercial, and advertisement of any type somehow manages to say outright or imply that "quality" is involved.

An industry and government requirement for "software" quality assurance has grown from a minimal force a decade ago into a major demand. The result has been the development of a whole new set of Quality tools and disciplines—a process that even today is still not complete.

Many companies, large and small, have adopted some of the total array of Quality tools or techniques or have even applied complete Quality System approaches to the conduct of their business. The federal government has begun to introduce "Total Quality Management" (TQM) into its day-to-day operations and has initiated the Malcolm Baldrige National Quality Award (MBNQA) to recognize "world class" accomplishments in the Quality field.

As a result of all this, the revisions and additions contained in this edition of *The Quality System* are substantial in scope and nature. Elements have been added in subsystems 2, 5, and 6 to address all of the above developments and other matters (including the logical change from "vendor" to "supplier"), and both Sections 10 and 11 (Quality System Engineering and Managing the Quality System) have been expanded.

It has been my intent to make the few additions here and there that would make *The Quality System* more directly translatable to both TQM and MBNQA usage (at least as the two of them have evolved by 1989). I have also tried to give more emphasis to the point that the Quality System could be directly applied to service industries, institutions, government agencies, and even the non-product-related functions within a manufacturing corporation. This does not mean that I have abandoned the primarily manufacturing-related examples and wording—just that I have amplified them in some areas to make the point.

In reviewing the material in the book, I find that I have used the word "program" to describe the scope of activity of one or more of the elements. It's a good word, but it is now looked upon with some disfavor. Wherever you find it, then, please mentally institutionalize what it reflects as a perpetually ongoing effort of your Quality System.

With that explained, I would like to acknowledge again the contributions of good friends and family members in getting me through this difficult process—I hope to the extent that you, my customer, will find this edition even more useful than you have told me the first edition was.

My thanks go to Roger L. Lohn for his help with the software additions to Subsystem 5; to Timothy J. Guido of Peterbilt Motors Company for the discussion of risk and loss in 6.1 Classification of Characteristics; to my son, James A. Caplan, for his recollections of pertinent events; to the late and lamented Richard A. Freund, for his assistance in keeping the terminology straight; and to my beloved wife, Shirley, for her continuous encouragement and forbearance during the throes of composition.

Preface to the First Edition

The objective of a Quality System is to achieve complete customer satisfaction at appropriate benefit to the producer. (*Customer*, in this context, means each person's "internal" customers as well as those of the organization as a whole.) The purpose of this book is to provide the guidance that every organization needs and can use to reach that objective.

Although the wording in much of the book has a strong manufacturing orientation, practically all of it can be adapted by a resourceful person to service, government, or commercial activities—thus the use of the word "benefit" in the objective, rather than the word "profit." As another example, a *customer* for a hospital might be a "patient"; for a law firm, might be a "client"; etc.

This is not a book on statistical quality control; the techniques appropriate to that area are thoroughly covered in the voluminous literature on the subject. Rather, it is a *management*-oriented description of all the activities required to develop and operate a completely effective Quality System in any enterprise.

In the 1950s it became apparent that effective management was a critical ingredient for the successful control of quality, that quality control was not just a stable of statistically based techniques to improve products and processes. The concept was expounded in two books written by Dr. A. V. Feigenbaum, the second of which, *Total Quality Control* (McGraw-Hill, 1961), became the primary reference in the field. In his books, Dr. Feigenbaum laid the groundwork for the collaborative or "system" approach to achieving quality.

The American National Standards Institute (ANSI) Accredited Committee Z-1 on Quality Assurance has spent a great deal of time and effort developing three documents that apply to quality systems: ANSI/ASQC A1-1978, *Definitions, Symbols, Formulas and Tables for Control Charts*, ANSI/ASQC A3-1978, *Quality Systems Terminology* (A3), and ANSI Z1.15-1979, *Generic Guidelines for Quality Systems*. I participated as a reviewer in the preparation of the latter two documents. All terms used in this book are intended to be compatible with the first two standards. The book is fully supportive and an amplification of the *Generic Guidelines*, and a Quality System developed in accordance with the concepts described in this book should satisfy all requirements imposed by any government or external source.

A Quality System is, according to A3, "the collective plans, activities, and events that are provided to ensure that a product, process, or service will satisfy given needs." The same document recognizes that "quality" includes "economics, safety, availability, maintainability, reliability, design, and all other characteristics that the need for the product or service involves." It can readily be seen that nothing less than the achievement of complete customer satisfaction is intended, and that the Quality System involves the coordinated activities of every person in an enterprise.

The Quality System outlined here is strongly geared to preventing quality problems from happening at all. Improved profits and reliability, as well as other positive effects of prevention programs, will be seen only after they are in place and working, not at the moment of application. To realize optimal benefits, management must be prepared to make an investment of resources, especially people, in the early stages of introducing a Quality System. Management must ensure that the group involved in planning the Quality System includes people who have the necessary knowledge to develop the system documents to their full effectiveness. This is covered further in section 10, "Quality System Engineering."

The contents of *The Quality System: A Sourcebook for Managers and Engineers* is divided into nine subsystems. Each element of a subsystem covers everything an organization should consider when planning the program for that element. For those elements that should be a part of your Quality System, study the descriptions carefully and imaginatively. Make check lists in your own terms to ensure that every needed aspect of the subject is appropriately included.

Virtually nowhere do I identify *who* should do something, nor do I state categorically *how* it should be done. From a system viewpoint, what is important is that the work be done effectively and efficiently. Who does it and how it can best be done are matters influenced by the formal and informal structures of the enterprise, its "culture," its products, and the management styles of its key managers. One of the difficulties that severely hampered the early efforts of the statistical quality control movement was its practitioners' insistence that they had to do or oversee everything. This was never acceptable to the managers of other functions and, in the Quality System concept presented here, it is viewed as a negative approach. This may explain my avoidance of assigning such responsibilities. To achieve optimum results, each enterprise must design the particulars of its programs and assign the responsibilities for them to fit individual circumstances and needs.

A number of people assisted in the preparation of this manuscript through suggestion or review. I would like to recognize particularly the substantial contributions of Roger L. Lohn of Motorola and of my son, James A. Caplan. Both of them provided many useful suggestions for improving content, construction, and readability. Additionally, Richard A. Freund helped ensure that the terminology was compatible with the latest ANSI/ASQC standards and made several other helpful suggestions.

The QUALITY SYSTEM

Introduction

The Quality System

In the sense used in this book, a "system" exists when data are gathered and analyzed and this information is used to produce an intended result. People and various kinds of equipment play key roles in most systems, particularly in management systems, of which the Quality System is a major example.

An analogy might help clarify what a system is. Consider the system of conducting a national election. There is a subsystem involving selection of official candidates to represent their parties in the general election. Within that subsystem, activities are associated with primary elections, caucuses, other methods of identifying candidates, protecting the validity of the selection processes, qualification of voters, notification and certification of results, and with other important details. There is a subsystem involving the publication of the date and circumstances of the election(s). Another subsystem focuses on the personnel requirements; clerks, watchers, auditors, and others must be hired or selected and trained. Still another subsystem deals with the ballot or voting machine logistics, their preparation, provision in adequate quantities, safeguarding, handling of absentee voting, reporting of results, and so on. All these subsystems and elements are involved in the electoral process. All work together within the system to produce a valid election.

Determining the structure of any system involves identifying those activities that must be performed to ensure that *all* the necessary information is developed and used effectively to achieve the planned result. That planned result, for most Quality Systems, is to achieve complete customer satisfaction at appropriate benefit to the producer. Customer satisfaction, in this context, includes that of each person's "internal" customers as well as those of the organization as a whole. The customer array must be identified and addressed for each individual and group. Most systems of any importance involve a considerable number of activities, requiring organization into hierarchies of subsystems and elements. The subsystems normally specify those activities that have some common orientation; the elements are the statements of the activities themselves.

Applying the System

The Quality System described in this book is divided into nine subsystems:

1. Quality System Management
2. Product Development Control
3. Purchased Material Control
4. Process Development and Operation Control
5. Quality Data Programs
6. Special Studies
7. Quality Measurement and Control Equipment
8. Human Resource Involvement
9. Customer Contact

Each subsystem contains a number of elements. I have attempted to make each element complete in itself, covering the planning and the execution of that plan in one statement, even though planning is often done by organizational components different from those who carry out the plan. This split of responsibility can be accommodated readily in documents and procedures like that shown in Appendix A. In five cases the planning operation is covered separately, usually because the development of the plan is, in itself, a major accomplishment.

The element-by-element Quality System is designed to enable the manager to build a more efficient operation at less cost. In building a Quality System, it is unlikely that any one enterprise would find every one of the 109 elements equally pertinent or even necessary. This is especially true for small operations. An attempt has been made, therefore, to highlight those elements most likely to be critical to most organizations. This is done through the establishing of priorities, as described in section 10, Quality System Engineering, and, more specifically, in the subsystem evaluation process presented in section 11, Managing the Quality System.

Whatever elements comprise it, though, the Quality System, like any other system, achieves its objectives through the performance of mutually supportive activities. One of the major problems of most nonsystematic approaches to quality control is their failure to recognize the interrelatedness of all activities. In innumerable cases, perfectly good statistical quality control and other beneficial techniques have been applied, used effectively for a while, and then abandoned for some reason. (See 6.8 Statistical Technique Application for further discussion.)

People try to apply single elements and become discouraged when no miracles occur. Some years ago, many companies, having heard good things about the Ford Motor Company's Vendor Certification Program, decided to adopt the concept as a way of reducing incoming inspection costs. Unfortunately, they neglected the other elements of Subsystem 3, Purchased Material Control, which Ford was using, and so their programs failed miserably, often catastrophically. These failures were blamed on the program, of course, not on the companies' lack of awareness of Quality System considerations.

While products and processes have become more complex over the years, customer demands have been growing stronger. Consumerism, environmental considerations, and government regulation have imposed new requirements on industry. Competition, particularly from foreign countries, challenges producers, especially in the quality area. These combined pressures have forced manufacturers and other potential users of the Quality System approach to expend more and more effort and money in this area.

For many companies quality costs have risen from a tiny fraction of sales to significantly more than 10 percent of sales. (I know of some cases where quality costs have exceeded 50 percent of sales over long periods of time.) This has produced, for many, a significant profit erosion or, at least, a failure to achieve attainable profits. *In many cases, the potential increased profits from the application of a well-planned Quality System are considerably greater than those realizable from doubling the size of the enterprise at current profit levels*—which are typically 4 to 5 percent of sales.

Additionally, increasingly demanding customers often view the attained quality level as "worse then ever" when, in fact, it may be substantially better than ever on an absolute basis. Consequently, companies that are not using a Quality System approach to satisfy their need to improve quality are paying more and yet achieving less in the critical field of customer satisfaction. Some have been forced out of business because of losses, sometimes as the result of a single product liability judgment or of a single regulatory action. The Quality System will not automatically prevent all such disasters, but will reduce the probability of disaster to a very low level.

In the case of manufacturers whose products are regulated by law or who are suppliers to government agencies, to prime contractors to the government, or to certain other large customers, freedom as to the form and extent of their Quality Systems may be curtailed. Contracts and regulations frequently intrude upon supplier decision processes, limiting the actions that may be taken. This situation must be recognized and dealt with from the beginning. Ground rules and limits must be established; the programs necessary to work effectively within the scope of the conditions must be prepared; and, where the imposed conditions are not realistic or not capable of being met economically, alternatives consistent with the intent of the original requirement must be developed and approved.

Such a situation may cause major changes or additions to an existing Quality System, and those changes may require further adjustments in the supportive elements. But such a situation also permits the full realization of the utility of the Quality System in coping with new circumstances. Properly designed and managed, the Quality System has the flexibility needed to adapt effectively to changing conditions.

When the customer is a government agency, the supplier must consider not only the contract terms in adapting the Quality System but also the subsidiary or "reference" requirements, as well. Many companies have found it vital to prepare a "specification tree" that covers not only the specific requirements of the contract, the drawings, and listed specifications, but also the specifications referred to in the listed items, the additional specifications referred to in those specifications, and so on. A small manufacturer of my acquaintance was forced into bankruptcy because of a heat-treating requirement in a "second tier" specification pertaining to his contract with a government agency. The specification containing the requirement was neither listed on his contract (it was incorporated by reference in a specification listed on a part drawing), nor was it presented to him by anyone until he had processed the entire order beyond the point of rework. He had assumed—to his great loss—that he was responsible only for the requirements directly covered by his contract.

In recent years, service organizations and the non-product-related departments of manufacturing companies have applied Quality System concepts to their operations. By thus recognizing that *all* functions in any enterprise must concern themselves with the quality implications of their work, these organizations are able to achieve major improvements in both customer satisfaction and their own operating efficiencies. These activities have, as a result, broadened the scope of such elements as 1.7 Quality Cost Program; 1.12 Corrective Action Program; 1.14 Audit of Procedures, Processes, and Product; and 6.6 Problem Solving.

Program Management

Just about any enterprise that has been involved with government contracts, and many other large enterprises, too, will be familiar with program management, a method of directing a complex undertaking requiring coordinated effort from numerous groups or manufacturers. Any large contractor must ensure the continuing utility of his Quality System by applying program management concepts to the specific contracts to be completed. Furthermore, the development or major modification of any Quality System involves extensive coordination of activities in a manner directly akin to program management. A review of the established principles of program management is therefore in order. The following outline headings are adapted from several government publications on the subject.

A. Establishment of program objectives

1. Negotiation and clarification to ensure understanding
2. Reconciliation with fundamental enterprise objectives
3. Separation into functionally oriented subobjectives for planning
4. Application of management approval controls

B. Resource identification for program development

1. Estimation and scheduling
2. Determination of availability and substitution patterns
3. Obtaining of resource commitments
4. Application of management approval controls to budgets

C. Program planning

1. Milestone identification and scheduling
2. Application of planning techniques (e.g., PERT)
3. Negotiation and assignment of major responsibilities
4. Feasibility review and risk assessment
5. Preparation of contingency plans
6. Planning for progress and results measurement

7. Planning for information flow and documentation
8. Selection of operational controls and establishment of standards
9. Planning and negotiating for limited program tryout
10. Application of management approval controls to plans

D. Program development

1. Operation of all program plans, including contingency (as needed)
2. Management by exception against milestone schedule and results achievement
3. Completion of program development subobjectives
4. Application of management approval controls to proceed to tryout

E. Program tryout

1. Training of involved personnel
2. Activation of responsibilities
3. Review of controls in operation
4. Identification of problem areas and institution of required corrective actions
5. Ensuring the effectiveness of changes
6. Operating resource determination and scheduling
7. Application of management approval controls to place program into full operation

F. Program operation

1. Resource application
2. Progress evaluation and program results measurement
3. Information flow effectiveness determination
4. Audit of procedure conformance
5. Corrective action and management follow-up
6. Program modification control

Steps A through E in the program management outline apply to the activities described under section 10, Quality System Engineering. Step F pertains to section 11, Managing the Quality System. Those sections may be more readily appreciated in the context of the program management principles stated above.

Subsystem 1

Quality System Management

The Quality System Management subsystem provides the information needed for the ongoing management of the entire quality system, ensuring its continued applicability and effectiveness. Within this subsystem are four elements that heavily involve activities of two or more other subsystems and cannot therefore logically be assigned to any one of them. These four elements are 1.2 Quality Standards (which involves Subsystems 2, 3, 4, and 9), 1.9 Material Tracking (which consists of activities associated with Subsystems 2, 3, 4, 5, and 9), 1.11 Product Certification (which involve Subsystems 2 and 4), and 1.13 Product and Process Modification.

The remaining thirteen elements address directly the broadly applicable support and control programs that are critical to the success of the Quality System. Included are such items as the overall program for auditing the Quality System (1.14); the Corrective Action Program (1.12); the Quality Cost Program (1.7); the establishment of Quality Policy (1.1) and Quality Objectives (1.3), assured of accomplishment through Quality Program Effectiveness Review (1.15); and the setting of Quality Standards (1.2). These programs, coupled with the rest of the elements in this subsystem and the more restricted controls of Subsystems 2, 6, and 9, provide the management activities and visible results measurement devices that assure the effective operation of the entire Quality System. The elements of this subsystem are arranged in the order of their likely development in an existing organization.

1.1 Quality Policy

Quality Policy covers the preparation, issuance, and control of relevant, formal statements used as the bases for all quality-related activity. It is often difficult to decide what to include and what to exclude in published policy statements. To begin with, policy in any area is the aggregate of management decisions for all questions associated with that area. Some such decisions, though, are applicable for such a brief period that their formal documentation is not justified. Others may be too insignificant or lacking in sufficient breadth of application to warrant documentation. In any event, Quality Policy should address questions of customer satisfaction and product- and service-quality improvement.

Most persons agree that there should be a one-page broad statement of Quality Policy signed by the top executive of the organization—the president of the company or the general manager of the division, for example. Such a statement provides direction for the entire organization. Consider, for example, a statement to this effect: "The products of this company will be of consistently higher quality, reliability, and safety than those of all our competitors." It lays an entirely different charge on the company than this one: "The products of this company will have at least as high quality, reliability, and safety as those of our major competitors." The wording is very similar. The differences, though, translated into objectives and specific goals, have both short- and long-range implications of major importance to both organizations. Yet the policy statements are both reasonable aspirations and in neither case specify anything exactly that anyone must do.

More specific direction is provided through a chain of documentation derived from the broad policy statement. In many organizations, that chain involves three additional levels of documents, having numerous forms.

The second level of documents (the statement of Quality Policy is considered the first) is often also called Quality Policy, since it still addresses its subjects in general terms or affects only a small portion of the Quality System. Here, decisions are made for the ultimate shape of the Quality System (for example, "Product design reviews involving all interested functions will be held throughout the design cycle" and "Inspection and test of product during manufacture will be performed by personnel reporting to Production"). Of course, formal documentation of these statements and their ramifications would involve more words than given here.

This leads to the third level of documents—often called "procedures." As might be expected, it is often difficult to separate the second level from the third. Perhaps the simplest way is to state that "procedures" tell how to carry out "policy," who does it, and under what circumstances and timing. A procedure might explain how Production would plan, specify, schedule, and control the inspections and tests they perform and how the Quality function would similarly accomplish its audits of those Production activities. For an example of a procedure, see Appendix A.

The fourth level is the extremely specific, detailed statement of work, often called "instructions." Here we might give the exact "how to" of the Production test of a certain part or assembly, listing the equipment and fixtures used, the test steps in sequence, the results reporting, and so on. An example of an instruction is shown in Appendix B.

As stated in the beginning of this element description, we are concerned here with the preparation and control of Quality Policy—therefore, with just the first and second level of documents. The statement, composed of a group of single documents for each organizational component, is not difficult to identify and incorporate into the program. The second level of documents could be voluminous, so selection criteria must be established.

Criteria for deciding which management decisions should be published as Quality Policy might be one or a combination of the following:

Publish only decisions affecting two or more departments.
Publish only decisions made by managers who are above level ____.
Publish only decisions affecting Quality System procedures.

Whatever criteria are adopted, they should be properly publicized and they should establish mechanisms whereby decisions meeting the criteria are submitted for review and publication. It is often desirable to have Quality Policy published on distinctive paper, collected in binders, and used as a comprehensive guide to decision making. By this means, everyone involved can be readily directed into compatible actions, aimed to achieve the Quality Objectives (1.3) of the organization.

Policy should be subject to regular review and modification as circumstances and responsibilities change. Since these documents reflect management decisions, the review and modification should involve upper levels of management. This ensures that both levels of Quality Policy will become and remain compatible with each other and with the real intent of management in the critical areas of the Quality System.

Quality Policy is the foundation for all activities of the Quality System giving rise to 1.3 Quality Objectives and associated goals. As such, it must be carefully developed and controlled to ensure the realization of the benefits of the System.

1.2 Quality Standards

Several different types of standards are referred to in the Quality System. There are documents called standards—developed by governments, industry groups, or individual companies—that are intended for broad application and full acceptance. Visual and physical standards, which also may have a variety of sources, are used as the basis for comparison with inspection or test results for judging conformance. And there are also performance standards, time standards, cost standards, and workmanship standards, among others, that apply to various aspects of quality or of the Quality System. The latter group also include those to be used by non-product-related functions such as finance, legal, and human resources.

All of these standards have the common ingredient their name implies. They reflect what *should* be, the criterion against which operating results are compared. Therefore, it is necessary, but sometimes difficult, to define and develop unambiguous, precise standards that accurately reflect what is required. Many times the only feasible way to do this is through a consensus of experts. Also, it is difficult to keep standards up to date and unaffected by physical or operational deterioration. This is particularly true of the visual and physical models used regularly in a factory for training in the use of standards, to resolve inspection and test questions, or as the basis for setting up a process or machine.

Replaceable photographs, duplicate standards embedded in clear plastic, and standards translated into terms of immutable physical constants are some of the means that have been used to solve the problems of supplier, factory, and field degradation of visual and physical standards. There is no universally applicable method. The important thing is to recognize that such items are not physically permanent and that protective measures must be taken depending on the need and potential risk.

The use of standards in training often requires a broader array of samples than would be needed for "go/not go" borderline decisions, and for this purpose an assortment should be developed with care to avoid possible confusion of the trainee. A similar array can be used for checking inspector accuracy.

When providing inspection standards, it is important to specify the circumstances of their use—lighting, magnification, handling, and applicability under special or limiting conditions. For example, a standard for gravestone quality defines several types of nonconforming conditions, but then adds, "Anything invisible from three feet away is not a defect." It is also important to specify maintenance and updating practices, protection to be applied, and storage preparation and conditions. Where practicable, an end-use date should be given and disposal specified.

Field activities such as installation, erection, and service, as well as factory operations, lend themselves to the establishment of time and cost standards. When coupled with the relevant quality standards, they provide a way of measuring the effectiveness of their operations to managers of what are often unsupervised activities. To minimize the destructive effects of some incentive programs, the proper use of the Quality Standards, including workmanship, as part of the incentive measurement is highly effective.

Standards, whether written or hardware, must be valid. Therefore, in addition to being concerned with their physical condition, we must review them regularly to assure ourselves that they are still pertinent. Design changes, competition, new customer or regulatory requirements, new factory or field processes can all affect the validity or completeness of coverage of the standards. Sometimes, these conditions can develop without obvious ties to existing standards, although the program should provide for timely revision of standards with any related change. The regular, often annual, review closes the gap. Wherever practicable, employees using these standards should be involved in developing them and in reviewing them for continuing validity.

Quality Standards are the bases for decisions on whether parts, products, processes, and services meet requirements. The entire standards program is oriented toward achieving uniformity of action and decision, when these are useful to the company or to the customer. Such standards should be available wherever possible errors can occur through ambiguous or inconsistent direction to personnel, suppliers, and even industrial customers. The savings in costs and customer dissatisfaction by avoiding such errors should be obvious.

1.3 Quality Objectives

Quality Objectives are derived from Quality Policy. Essentially, the Objectives restate the Policy in specific terms. An example might be the following:

Quality Policy: We will produce a product of a reliability superior to that of our competitors.

Quality Objectives:

A. By five years from date, our field failure rate from all causes for product less than ten years old will be reduced to 20 percent of its current level.
B. By five years from date, our external failure costs will be reduced to 40 percent of their current level.
C. By two years from date, all new products will have certified reliability predictions, compatible with the above objectives, generated no later than three months prior to pilot run.

Such examples lead logically to the establishment of quality goals, as described in 1.15 Quality Program Effectiveness Review. The goals are the milestones that lead to the accomplishment of the Quality Objectives. Goals associated with the above objectives might be as follows:

	Goals				
Objectives	*This Year*	*2nd Year*	*3rd Year*	*4th Year*	*5th Year*
A	0	−5%	−25%	−50%	−80%
B	0	+2%	−10%	−30%	−60%
C	70%	100%	100%	100%	100%

Through the setting of Quality Objectives and goals, based upon the management commitments expressed as Quality Policy statements, the organization can concentrate on essentials, develop specific programs to achieve the desired results, have a visible and agreed-upon timetable for their accomplishment, and measure progress toward success. The goals can be built into management incentive programs, with appropriate acknowledgment of reaching a formal objective. All such objectives and goals must be compatible with each other and with all other objectives and goals of the organization. In addition, insofar as they are related to each other, they must be mutually supportive.

As a result of the establishment of Quality Objectives, supplemented by quality goals and verified by 1.15 Quality Program Effectiveness Review and by 1.14 Audit of Procedures, Processes, and Product, top management can be satisfied that all parts of the organization will be cooperating to produce the desired results. As experience dictates, revision of the Quality Objectives and the concomitant goals may well be appropriate from time to time, even before the allotted time has expired, since the tendency is usually to be too conservative when forecasting cost reductions and time savings from a well-planned Quality System program.

1.4 Quality Manual

Some customers, notably government agencies and their prime contractors, require their suppliers by contract to develop and submit to them for approval or have available for review a formal Quality Manual. Such a manual usually comprises a brief statement by the top executive involved (see 1.1 Quality Policy), supplemented by all the organization's current quality-related procedures that support the execution of the Quality Policy. These procedures are then expanded through detailed specific work instructions, where appropriate.

The Quality Policy, its related procedures, and the detailed instructions may be bound together to form the Manual, or the procedures and instructions may merely be identified through lists filed with the policy statement. In any case, all of them are made readily available to the customer's representatives. Most important, copies must be immediately available to company personnel whose work necessitates their use, and they should be kept current.

The two types of instructions contained in the total Quality Manual are identified as procedures and instructions. At least one procedure is required for every Quality System subject, selected from the elements described in this book and deemed appropriate to the operations of the enterprise. A procedure usually tells what must be done, how to do it, who does it, and in what sequence. When specific guidance is needed—such as which dimension to inspect, with what instrument, on which particular part—a work instruction is required. There must be a procedure telling how to develop such an instruction, who develops it, how and when it is issued, and how it can be revised. Instructions, therefore, are subordinate to procedures, just as procedures are subordinate to policy, and are issued only for situations that specifically require them. (See 4.4 Quality Check Stations.)

Procedures and work instructions should be issued in a consistent format, using the fewest words possible to transfer needed information. Each document should contain only the specific information needed by the intended user. A sample procedure is shown in Appendix A, illustrating all the useful sections. Appendix B is an example of an effective work instruction.

A Quality Manual as described above becomes a training aid for new and transferred or promoted personnel, provides the basis for resolving disputes rapidly, and serves as the reservoir for the evolved disciplines or "culture" of a growing organization. It also satisfies any contractual requirements for such a document and provides the standard against which to conduct both external and internal Quality System audits. (See 1.14 Audit of Procedures, Processes, and Product.)

1.5 Documentation Control

Although the following discussion is brief, the subject is of paramount importance, and ignoring this element will lead to chaos. Every producer must know accurately the evolutionary state of each item in production, in distribution, and in the customers' hands. This is the whole point of Documentation Control, and its ramifications are many.

The current version of all relevant documents and standards (and every producer can compile a substantial list of these) must be used by all involved personnel, and the version used at any given time must be known. The materials, tools, equipment, parts, components, subassemblies, assemblies, finished product, and anything else that can affect ultimate customer satisfaction used or applied at any given time must also be known. (This concept is frequently referred to as *Configuration Control* and is dealt with at greater length in 1.9 Material Tracking.)

Here is one example of what can happen when insufficient control of documents is coupled with other failures in management of the Quality System. A supplier of electronic subassemblies was not given the latest drawing revision, which involved only a change in the value of one component. He continued to obtain and supply the obsolete component in thousands of subassemblies for almost a year, and since the supplier was a subsidiary of the purchasing company, no incoming inspection was performed on his shipments. The incorrect components were replaced out on the line with no word to proper levels of management or to the supplier. The situation was uncovered only by an audit in the early stages of development of a full-fledged Quality System.

It cannot, therefore, be assumed that document and practice correspond. Many instances have occurred where the existence of a revised drawing or test instruction has been unknown to the people who should have been using it, or the revision has been ignored because of supplies of old parts or lack of revised test fixtures. Documentation Control involves the application of programs to ensure that proper knowledge exists at the right levels of producer management to prevent these and similar discrepancies.

Documentation Control is oriented primarily to detailed records, to the control of obsolete documents, to other paper work, and to reference standards. Yet it is important to recognize the necessary relationship of Documentation Control to control of material and equipment. As indicated above, the material control aspect of this subject is described in element 1.9.

1.6 Growth Planning and Control

Virtually every organization that provides a product or service does so with the conscious or unconscious expectation of growth. Such growth may be expected to be rapid or slow; to be accomplished by increasing acceptance in the marketplace, or by acquisition or other means; and to be accompanied by either stable or improved quality. But many times the actual results are at variance with the anticipated ones, and often the variance produces traumatic effects—particularly in the area of quality—with consequent increased costs and delays in schedule. This situation becomes even more critical when growth is negative rather than positive.

If a plan contains the correct ingredients and is properly carried out, either positive or negative growth can be achieved without sacrificing quality. In fact, if growth involves new or refurbished equipment and an opportunity to train people more effectively, quality can be significantly improved. This happens much too infrequently because quality improvement is not included as a primary consideration of the growth plan. The Growth Planning and Control element emphasizes improvement with growth.

The plan must provide for the elimination of sources of quality problems already experienced in producing the same or similar products. Among other activities, this may involve product redesign, new or changed processes, training of operators and others, deliberate changes in facilities, and establishment of new controls on processes. If the growth is taking place at a new and distant location, the plan must be concerned with the effects of the new environment on the ability of the processes to produce a satisfactory quality product, as well as with the effects of the processes on the new environment. New suppliers, transportation facilities, environmental conditions, personnel cultures, and ratios of labor versus equipment costs may all be involved in a new location situation. Their effects must be recognized, factored into the plan, and, if potentially deleterious to quality, compensated for or overcome. It is appropriate to perform a careful Process Design Review (4.1) to make certain that the transfer of existing processes will work. For example, water quality may be different, necessitating a major change in a plating process.

Even when growth takes place at the original location, some or all of these factors may adversely affect quality when increased production starts. More companies fail to achieve good quality results on start-up under these circumstances than when beginning production at a remote location. They assume that the introduction of a new production line in an existing facility will automatically result in a transfer of skills. An existing line is often essentially duplicated, with no thought to quality improvement or to the training needs of the new or transferred personnel. The result is often a quality disaster. The same thing occurs when reductions in volume and personnel involve "bumping down" without adequate training.

An increase or decrease in production capacity, whether local or remote, requires the preparation of a sufficient plan that will at least maintain the prior level of quality. Such a program will ensure realization of quality requirements (and often cost requirements, as well) on rearrangement, enlargement, reduction, or remote site location of manufacturing or service facilities.

1.7 Quality Cost Program

The Quality Cost Program covers the measurement, reporting, and use for resource allocation decisions of the costs incurred in ensuring product quality, reliability, and safety and the costs incurred from failure to do so—internally, at suppliers, and in the field. Although quality costs have been measured by some companies for many years as a device to support their quality control efforts through an identification of the real impact of those efforts, major application of quality cost programs in industry was instigated by the publication in 1963 of MIL-Q-9858A, the top-level quality document issued by the government for military contractors. Since then, many constructive efforts have been made to present this subject in a form that everyone can use.

Of the six measurements of effectiveness and control of the Quality System (see page 25), the Quality Cost Program is the most broadly applicable. This is because it communicates in terms readily understood by personnel at all levels within an organization—money. Obviously, the measurements and reporting must be timely, readily understood, and suited to the needs of management personnel, but the impact of the magnitude of these costs on profits as well as on departmental budgets and cost control objectives can be appreciated readily by all. For all managers to see the cost reduction effects that result from investments in prevention or appraisal, it is important to develop the program in such a way that it covers all significant quality costs and clearly distinguishes input from output.

Functions such as Accounting, Employee Relations, Security, Management Information Systems, Traffic, and others that are traditionally thought to have little or no involvement in actions counted as Quality Costs are beginning to realize that they have customers (often "internal" rather than "external") and measurable quality of their efforts (thus generating Quality Costs). Government agencies, schools, hospitals, banks, insurance companies, and law firms are examples of service groups that can also benefit from the application of Quality System concepts and the associated management of Quality Costs. I have included examples of some related cost accounts in the Chart of Quality Cost Accounts, which follows.

As an organization begins to invest in a formal Quality System, the establishment of a good quality cost measurement program becomes essential; it gives needed visibility to the results of corrective action (see 1.12

Corrective Action Program) and prevention activities and provides guidance to management for the movement of funds from the "failure" categories into further "prevention" investment.

Many of the quality cost programs developed in industry have erred in not showing this distinction more clearly by failing to adhere to the message implicit in the definition by Feigenbaum in *Total Quality Control* (McGraw-Hill, 1961, p. 83): "Prevention costs are for the purpose of keeping defects from occurring in the first place." This concept cleanly separates the inputs from the outputs. When prevention works, there are very few failures, if any. Thus the corrective actions of product redesign or process modification and their attendant costs become small or unnecessary. Corrective activities cloud the arithmetic when they are mistakenly viewed as preventive in nature (in the sense that they "prevent" more nonconformities from being generated). As we increase prevention effort through the application of the relevant elements of the Quality System (e.g., 1.14 Audit of Procedures, Processes, and Product, 2.9 Product Design Review, 4.1 Process Design Review, 6.1 Classification of Characteristics, and 8.10 Personnel Quality Participation, among others), the result should be seen in reduced failure costs. When corrective action programs are viewed as prevention, part of the reduced costs due to corrective action will be in the prevention category, offsetting the planned prevention program increase and confusing everyone.

Chart of Quality Cost Accounts

It is highly desirable to develop the chart of quality cost accounts to show changes as they occur, with deliberate inputs separated from their results. The list of the components of a thorough Quality Cost Program chart of accounts follows, divided among the four usually encountered categories of prevention, appraisal, internal failure, and external failure. It is customary for the costs to include noncapitalized wages, salaries, materials, rentals, outside agency charges, other expenses, and appropriate portions of overhead, whether reimbursed by suppliers or customers or offset by sales (of scrap, for example). The chart is flexible; the items in it may be individually used or discarded in the program for any one enterprise.

In addition, there may be situations where it is impracticable or overly expensive to collect hard monetary costs for large contributors to the total Quality Costs. When such a condition exists, it may be useful just to count the events and then apply a carefully developed monetary value or formula to the total number of these events during the period under study.

A. Prevention (the costs of those Quality System activities that are meant to keep nonconformances or customer dissatisfaction from occurring in the first place)

1. Costs of planning and administering the Quality System

a. Developing and maintaining the Quality System in all its aspects

(1) Internal costs
(2) Consulting and other external agency costs

b. Performing all those activities that ensure that the products and processes are wholly compatible with each other and suitable for control

(1) Determining that customer requirements match the organization's capabilities
(2) Performing necessary studies to support the quality and reliability portions of bid proposals
(3) Performing the quality engineering effort required for new product and service designs, including tolerance evaluation, reliability prediction, inspection and test requirement determination, yield and accuracy forecasting, and so on
(4) Performing the quality engineering effort required for new process designs, including process capability, yield forecasting, inspection and test requirement determination, and so on
(5) Performing the quality engineering required for new service offerings, including making compatibility checks with existing offerings, developing pertinent effectiveness measurements, ensuring that related training and information programs are appropriate, and so on

c. Developing the quality and reliability plans associated with a new or changed product, process, service, or Quality System program

(1) Preparing the specific quality and reliability plans for all aspects of the new situation, including supplier controls, process controls, finished product controls, installation and service controls, and management controls

(2) Developing the necessary quality and reliability specifications, procedures, work instructions, manuals, and standards to support the new item

d. Determining the cost of quality and reliability assurance and control effort associated with new or changed products, processes, services, or Quality System programs

(1) Performing such cost analyses
(2) Developing a quality cost accounting program

2. Costs of product design control engineering

a. Reviewing product and service designs for probability of achievement of relevant performance, yield, reliability, and safety goals; manufacturability; serviceability; and tolerance and interface compatibility among all characteristics, including packaging of the product

b. Performing necessary inspections and tests and reviewing the results for demonstrated achievement of all product goals in prototype and pilot run evaluations

c. Re-performing critical engineering calculations and measurements, including exploring alternate design evaluation or approaches

d. Checking drawings and other specifications to eliminate errors

3. Costs of process control engineering

a. Reviewing internal process designs and actual performance for inherent capability to meet quality and safety goals and to provide quality certification

b. Planning and establishing appropriate operating controls, such as control charts for manufacturing, service, and non-product-related processes; sampling plans; and receiving, setup, first piece, and patrol inspection

c. Determining suppliers' apparent and actual capability to produce items of requisite quality and cost

(1) Planning and performing pre-award surveys and performance analyses

   (2) Planning and conducting formal rating of the supplier's quality, reliability, delivery, and related performance

   d. Working with suppliers to ensure their receipt and understanding of proper quality requirements and their application of appropriate Quality System prevention-oriented programs, including packaging the items, and exhibiting effective concern for our personnel's safety in handling the purchased material

4. Costs of providing quality measurement and control equipment

   a. Designing and constructing or procuring conceptually new (to us) devices for inspection, test, and manufacturing equipment control

5. Costs of providing a capable and dedicated work force

   a. Determining and satisfying the personnel quality, safety, and related technical and management qualification needs of the entire organization through testing, training, and certification programs internally or externally developed and administered
   b. Developing and operating programs to obtain full participation of all personnel in preventing and correcting quality and safety problems

6. Costs of assuring the successful operation of the Quality System

   a. Auditing and reporting on the organization's compliance with requirements of the Quality System programs, procedures, instructions, specifications, and standards

   b. Analyzing all Quality System data supporting the control mechanisms, as follows:

      (1) Quality Cost Program (1.7)
      (2) Audit of Procedures, Processes, and Product (1.14)
      (3) Customer-Centered Quality Audit (2.10)
      (4) Yield Control Program (4.13)
      (5) Field Problem Controls (9.12)
      (6) Customer Satisfaction Measurement (9.14)

   c. Providing preventive maintenance as required for all equipment used to produce or appraise product or service

7. Other prevention costs

   a. Providing for administrative costs of Quality System activities not covered elsewhere

   b. Performing problem-solving activities unrelated to manufacturing processes or product-related considerations

   c. Providing for extra supplier charges for doing any part of this prevention work

   d. Conducting continuous quality improvement projects after the process capability meets or becomes better than the corresponding tolerance limits

B. Appraisal (the cost of inspection and test, other than developmental, of the products, processes, and all their constituent or replacement parts)

1. Costs of preparation for inspection and test

   a. Providing proven, established, or "off the shelf" equipment for inspection and test through

      (1) Purchase
      (2) Internal manufacture
      (3) Rental

   b. Setting up equipment to perform inspection

   c. Setting up equipment to perform test

   d. Providing consumables for test

      (1) Products
      (2) Utilities
      (3) Supplies

2. Costs of purchased material control (see C. Internal failure, 3 and 4a)

   a. Performing (or witnessing) and reporting the results of inspections and tests of purchased direct and indirect material and equipment at the supplier's or an intermediate location, including laboratories

   b. Performing and reporting the results of internal inspections and tests of purchased direct and indirect material and equipment, including laboratory testing

   c. Performing and reporting the results of audit of supplier operation and product, including packaging and preservation

3. Costs of internally produced material control (see C. Internal failure, 2 and 4a)

   a. Performing and reporting (including posting data on and analyzing indications of control charts) the results of process control inspections, tests, and surveillance of equipment, tooling, fixturing, and process parameters, including special processes

   b. Performing and reporting the results of material acceptance inspections and tests on parts, components, subassemblies, assemblies, and final product during production

   c. Performing and reporting the results of material acceptance inspections and tests on finished product, including packaging

   d. Performing and reporting the results of reliability and safety testing of the product

   e. Performing and reporting the results of audit of finished accepted product, including after packing

4. Costs of control laboratory operation (see C. Internal failure 2, 3, and 4a)

   a. Performing and reporting the results of calibration and maintenance for all quality measurement equipment, for production equipment used as a medium of inspection, and for quality standards

   b. Performing and reporting the results of process control analyses of chemical and metallurgical processes and product during processing

   c. Preparing and shipping product, with its associated records, to outside laboratories and equipment (with its records) to laboratories or storage

   d. Evaluating and maintaining the continuing utility of items in storage

5. Costs of control over service and non-product-related processes (see C. Internal failure 5d.)

    a. Performing and reporting (including posting data on and analyzing indications of control charts) the results of process control inspections and tests

6. Costs of approval by outside organizations

    a. Obtaining approval from necessary sources, such as

        (1) Government agencies
        (2) Underwriters Laboratories
        (3) Insurance underwriters

7. Costs of releasing product for production and shipment through analyses of inspection and test data

    a. Determining and reporting adequacy of new product design as demonstrated by inspection and test performed during development

    b. Determining and reporting conformance of production product to appropriate acceptance and customer criteria during processing, for release to next operation, and prior to release for shipment

8. Costs of field quality demonstration

    a. Determining and reporting the quality of parts, assemblies, or final product in distribution or at the customer's location prior to acceptance

9. Other appraisal costs

    a. Providing for extra supplier charges for doing any part of this appraisal work

    b. Accumulating data processing costs associated with the Quality System

C. Internal failure (the costs, incurred before receipt of the product by the customer, of all activities necessitated by the failure of suppliers or of enterprise personnel or equipment to produce a satisfactory product)

1. Costs of scrap
   a. Investigating, processing, replacing, and accounting for nonconforming material that cannot be used for any production or product purposes
   b. Costs of the material scrapped, including value added to the point of scrapping
2. Costs of rework and repair
   a. Investigating, processing, performing repeat or omitted manufacturing operations on, and accounting for nonconforming material that can be restored to original specification condition
   b. Investigating, processing, performing manufacturing operations on, and accounting for nonconforming material that cannot be restored to original specification condition but can be repaired—thus making it capable of being used for its original purpose
3. Costs of supplier-caused scrap and rework
   a. Investigating, notifying supplier, negotiating, processing, returning to supplier or to some other outside agency, performing repeat or omitted manufacturing operations on, cancelling orders for, and accounting for nonconforming purchased material that, because of supplier error, may or may not be able to be made usable
   b. Costs of any material scrapped because of supplier error, including value added to the point of scrapping
4. Costs of utilizing nonconforming material for other purposes
   a. Downgrading the product to a lower-quality classification, with attendant reduction in income, or converting the product to some purpose other than originally intended, with additional manufacturing cost (salvage)
5. Costs of corrective action from nonconforming material and processes
   a. Performing nonconforming product disposition effort, including material review activity

b. Performing failure analysis of purchased and produced nonconforming material to determine causes

c. Developing and testing corrective action solutions, including those necessary for permanently eliminating the cause(s) of nonconformities

d. Correcting procedures, policies, and instructions in any area where quality problems are identified—including such functions as accounting, personnel, legal, or other service or support activities

e. Correcting the product designs, specifications, the company and supplier processes, and involved manufacturing and measurement equipment to eliminate the causes of the nonconformity, both short-range and permanently; includes "maintenance of line" engineering

f. Sorting or reappraisal of nonconforming, reworked, or repaired material or equipment

g. Performing additional manufacturing or appraisal operations other than sorting beyond those scheduled, because of nonconforming material—including those thereafter made part of the routine direct manufacturing process

h. Following up on corrective action activities to determine their full effects

6. Other internal failure costs

a. Losing production time or productivity as a result of safety problems or of quality failures causing material lack or equipment problems

b. Losing quality or reliability incentive fees

c. Providing for extra supplier charges for failure analysis and other internal failure reactions

d. Replacing material lost or damaged in transit or from natural disasters

e. Carrying charges on money tied up in payable/receivable float on material returned to supplier

f. Carrying charges on that portion of inventory that failed to ship as planned because of quality problems

D. External failure (the costs of all activities resulting from failure to meet the product and service requirements of the customer to his satisfaction)

1. Costs of field failures in warranty

a. Investigating and analyzing field failures to identify needed corrective action for the failed items and other items in the field

b. Repairing the failed item and others like it as required and meanwhile providing appropriate support for the customer

c. Replacing the failed item or providing a monetary allowance to the aggrieved customer

d. Correcting the product designs, specifications, company and supplier processes, and involved manufacturing and measurement equipment to eliminate the causes of the nonconformity, both short-range and permanently

2. Costs of field failures out of warranty

a. Investigating and analyzing field failures to identify needed corrective action for the failed item and other production and negotiating with the customer about the necessary correction to the item

b. Repairing the failed item and others like it as required and meanwhile providing appropriate support for the customer

c. Replacing the failed item or providing a monetary concession to the customer

d. Correcting the product designs, specifications, company and supplier processes, and involved manufacturing and measurement equipment to eliminate the causes of the nonconformity, both short-range and permanently

3. Costs of Customer Complaints

    a. Maintaining a customer complaint agency or function to provide ready access and rapid, effective response for the customer
    b. Investigating and resolving any complaints against products or services received from customers
    c. Replacing the item or service or providing a monetary concession to the aggrieved customer when no other remedy is effective

4. Other costs of field service resulting from product or service problems

    a. Correcting nonconforming conditions, including any needed extra testing, not the result of a customer complaint, and meanwhile providing appropriate support for the customer
    b. Carrying charges on customer receivables not liquidated because of quality problems

5. Costs associated with product liability

    a. Providing for insurance against product liability
    b. Operating any product recall campaign
    c. Sustaining lawsuits or government regulation

Exposing Problem Areas

The external failure costs are the most important in the long run, since they reflect the customers' view of our product. If they are high, we are in danger of losing the customers' goodwill, even if we avoid devastating product liability suits or government punitive action. Our programs should be directed toward the *elimination* of field failures and their attendant cost. The old concept that this approach is unjustifiable economically is no longer valid—if it ever was. Too many external factors have entered the equation, particularly in matters affecting product safety, to permit anyone in the consumer product industries to conceive of "business as usual" as a satisfactory strategy.

Both appraisal and prevention efforts must be increased to force external failure costs down. Initially, this will increase internal failure costs as a result of increased detection of nonconformities and corresponding corrective action. But as the corrective action takes hold and as the longer-range prevention activities become effective, internal failure costs will also be reduced. Ultimately, appraisal costs can be reduced safely as nonconformities diminish.

The Quality Cost Program helps to pinpoint the areas where corrective action can have its most useful effects. This is accomplished by establishing quality cost budgets and goals for departments, products, suppliers, and any other useful category and then managing the operation so as to achieve those goals. When the goals are exceeded, either on an absolute or a statistical basis, corrective action is initiated. This calls for a sufficiently extensive investigation to determine the real cause(s) for the high cost figures (low, in the case of prevention costs), followed by the carrying out of the necessary programs to eliminate the cause(s). Should this prove to be ineffective, the Quality Cost Program would then call for the application of the provisions of the Corrective Action Program (1.12).

Quality cost problem investigations and cost reduction programs are initiated by excessive costs rather than by nonconformity rates, and they therefore require an analysis of the accounting data to locate the problem's physical source. Nevertheless, the mechanics of conducting these investigations and programs utilize the same techniques as those described under 6.7 Quality Level Improvement. It is thus possible that the same conditions requiring corrective action may be identified by indices other than quality costs. Therefore, the measurement programs of the Quality System must be carefully coordinated to supplement, not duplicate, one another.

This coordination requires careful cross-referencing of the quality cost accounting program with the data programs of the five other Quality System control devices listed on page 25. Whether these data programs are manual or computer based is immaterial; proper safeguards must be applied to prevent wasted effort. Each control program will initiate corrective actions, and the results will be most usefully reflected in improved quality costs, offering another reason for close coordination of the data programs.

Optimizing Quality Costs

There is a considerable body of literature covering the optimization of quality costs. A minimum (optimal) total cost point can be found on a curve representing dollars (costs) versus nonconformance level. The curve is a result of adding the cost-nonconformance curves of the prevention, appraisal, internal failure, and external failure cost categories. Lundvall and

Juran (*Quality Control Handbook,* 3d ed., McGraw-Hill, 1974) and Harrington (*Quality* magazine, May and June 1976) discuss this concept at length. Harrington adds customer-incurred costs and customer dissatisfaction costs to the four producer-oriented cost categories described above.

Certain customer costs, usually considered to be intangible as far as the producer is concerned, are often factored into the quality cost accounting program through a formula increase in observed external failure costs. The assumption is that any product deficiency seen by the customer that causes an action measurable in the producer's accounting system (e.g., investigation costs, warranty repairs, or out-of-warranty service) reflects other events that do not surface similarly. Some companies have multiplied the total measured external failure costs by a factor of 2 to as much as 10 in an attempt to represent the real impact of customer-observed quality problems. Other companies have tried to get more nearly exact measurements of these costs in individual cases to support their corrective action efforts.

Further justification for this approach is that many product failures never impact the producer's accounting system at all. Some customers repair items themselves, suffer with the deficiency, or just discard the items. Whatever happens, whether or not the event has tangible impact on the producer, there is a potentially even greater negative reaction by the customer, and this can result in his refusal to buy further items from the producer or even in his influencing others to refuse to buy the products.

The U.S. military and a number of corporations have introduced the concept of "Life-Cycle Costing" as a procurement contract device to force their suppliers to forecast the user's costs of owning a given product. Many of these costs are of the type considered here as quality costs. So a body of information and practice is developing that will ultimately cause the producer to consider these customer-ownership costs routinely as a portion of Quality Cost accounts. The current energy concern has focused attention on cost of ownership, although on a somewhat different basis, and this also has an impact on Quality Cost measurement. All these things contribute to the recognition that measured external failure costs represent only a small portion of the total impact of a potentially nonconforming product reaching the customer. This makes the Harrington approach cited above highly realistic today.

However, both the references cited, and many others, make an assumption in drawing their curves that I consider misleading. In all cases, the total curve of quality costs after reaching its optimum value is shown rising smoothly to an indefinite maximum, identified as infinity. This occurs because the sum of prevention and appraisal is shown doing the same thing. The assumption is that to achieve zero failure costs one would have to spend infinitely vast sums of money in prevention and appraisal. In practice, however, a product is created by a given process, and the quality costs are reduced by a variety of means, to a minimum. When the associated level of failures is deemed still unsatisfactory, particularly if further efforts to reduce it cause an increase in total costs, *a different process is applied.*

For example, one manufacturer's move from manual to mechanized production was accompanied by a drop in failure costs, appraisal costs, and total quality costs, as well as in total manufacturing costs. Despite this improvement, further prevention effort was applied in the design of the next level of production equipment capacity needed for the full output rate required. All in-process inspection was built in to the automatic production machinery, with control mechanisms included to ensure that no nonconformities were produced or, if produced, allowed to continue processing. The net result is that after production of some two million total items over a period of several years, measurable failure costs are infinitesimal. Although a few pieces have been scrapped and a very few have failed test, there has never been a lot failure requiring rejection, and there has never been a customer complaint or known field failure. Appraisal costs have been limited to a small amount of incoming inspection (much of it 100 percent automated), some process control by roving inspection, and final test (a standard sample per lot); all other appraisal is built into the automatic production equipment. Prevention costs, primarily invested in design review and in production and appraisal equipment design, were measurable initially but ceased with automatic production, except for occasional audits to ensure compliance with all requirements.

This pattern, which is by no means unique, shows that, if the quality cost curve reflected production processes, we should expect not one but *several* optimum points on a total quality cost curve for a product, depending upon the product's compatibility with the manufacturing processes, as we move from manual production to automation.

Even more important is that it is possible to achieve *zero* failure costs with minimal prevention and appraisal investments. Notice that this does not claim zero nonconformity levels, but zero failure costs. While some nonconformities and failures were observed with the example described above, the associated costs were so small as to be insignificant when viewed as an increment of production costs or of quality costs. Far from approaching infinity, the total quality costs in this example were so low that the total manufacturing costs were only about two-thirds the initial planning estimate—a remarkable accomplishment in the industry. The key to this is the deliberate forcing of the product design to be *wholly* compatible with the manufacturing processes to be used.

Reporting and Evaluating Quality Costs

The idea of looking at failure costs as an increment of production costs or of quality costs illustrates a necessary consideration of every Quality Cost Program: how to report the results meaningfully for each member of management. To be most useful, the cost program should recognize the areas of primary interest to various levels of management and of the different functions. The selection of the proper cost comparison bases can then give each member of management his most useful report. Most production managers will be interested in quality costs per labor hour, per unit of product, or as a percentage of production costs. Purchasing managers will be interested in supplier-related quality costs per dollar of direct material. Top management, Marketing, and Finance will relate best to quality costs as a percentage of net sales or of contributed value. Feigenbaum lists ten such comparison bases and discusses their relative utility under different circumstances. All managers will want to see their departmental quality costs, and those their departments cause, related to their budgets and to established goals. The reporting structure must address all of these needs, show trends, and highlight exceptions.

Once the usable quality cost elements have been identified for each function, the cost comparison bases have been mutually agreed upon, and the methods of data input have been determined, a user's manual should be prepared for each different situation. This could start with an identical chart of all quality cost accounts, selected from each functional grouping, and explained with examples relating to that group's activities so as to minimize ambiguity and reporting error. The manual should contain instructions on how to input the data, interpret the resultant reports, and handle corrective action signals.

With all this in place, the Quality Cost Program is ready to be put to use as the primary management measure of the effect of Quality System activities. It shows the improvements made by the Quality System in profitability, shows trends and pinpoints areas of additional profit opportunity (see Pyzdek, "Impact of Quality Cost Reduction on Profits," in *Quality Progress*, November, 1976, for a useful discussion of this subject), serves as the base for projecting costs with time and for new projects, and provides the basis for justifying investment for quality improvement in both capital equipment and other programs. When the operation of the Quality Cost Program or of other controls results in reduction in the level of nonconformities produced by a supplier, the data may be useful in obtaining reductions in purchase price of the material he provides.

The Quality Cost Program described here covers only a portion of all quality costs. It ignores the capital costs (floor space and capital equipment, including depreciation) and many of the indirect costs (suppliers' internal quality costs, overheads associated with producing nonconformities, productivity losses from quality problems, carrying charges on excessive and unmoving inventory and its storage, and customer cost resulting from produced nonconforming material). Were all these costs to be included, the enormous aggregate quality costs and the associated potential for cost reduction would be even more readily recognized.

A Start-Up List for Estimating Quality Costs

The list of quality cost accounts contains all of those activities likely to be of importance in one situation or another. Very few companies would find that *all* of them were of sufficient magnitude to be worth tracking. Thus, no two companies, even in the same industry (and, often, no two divisions of the same large company), would use exactly the same list of accounts. In addition, many of the items on the list may not be identifiable or practical to develop within a company's accounting system.

A rough estimator of quality costs is useful as a start-up device for any enterprise and perhaps permanently for small companies. It can be developed rapidly and has the virtue of reflecting the trends, if not the magnitude, of quality costs. A key characteristic is that it can be generated by the accounting department from those large items usually segregated in an existing accounting system. Even here there may be differences in what can be done in individual cases, but the list should include as many "real costs," not estimated, as practicable from the complete list given earlier. A typical selection might be as follows:

A. Prevention

1. All Quality department costs except those for inspection, test, and failure-related activities
2. Training in quality subjects
3. Preventive maintenance
4. New-product design checking and testing

B. Appraisal

1. All Quality, Production, and Service department inspection and test except reinspection and retest
2. Reliability testing of product
3. Control laboratory operations
4. Inspection, test, and laboratory expense equipment and supplies
5. Quality data generation, analysis, and reporting

C. Internal failure

1. Scrap
2. Rework and repair
3. Reinspection and retest
4. Sorting
5. Failure investigation and evaluation
6. Product and process design correction
7. Engineering changes
8. Purchase order changes

D. External failure

1. Warranty
2. Recall campaigns
3. Product liability insurance

Most of these accounts should be reasonably available in any accounting system, so it's a matter of having their total identified by the Accounting department on appropriate regular financial statements. The use of comparison bases, as discussed earlier, is possible to some extent even with this truncated approach. Certainly, viewing the total as a percentage of sales would be informative, at least.

Selling a Quality Cost Program to management is usually best accomplished by a special study of the pertinent costs from the twenty listed above. This study should be conducted by the Accounting department with guidance by the Quality function. Once the comptroller's office sees, from its own figures, how large a percentage of sales this portion of the total quality costs represents, it will usually cooperate actively or even take over the job of convincing top management of the need for such a program.

There is one mistake that large companies often make when comparing one division's quality costs to another's. Even with one centrally specified and audited accounting system, divisions often generate their accounting details differently. In addition, their products and processes may force major differences in their quality costs. This makes comparison of reported quality cost percentages misleading. Therefore, while the reported percentages should be examined with interest and investigated where indicated, only trends can be validly compared among divisions. Failure and total quality cost improvement as percentages of the previous year's average level (or some similar meaningful measure of change), not the reported quality cost numbers themselves, should be the corporate comparison measure.

The Quality Cost Program provides a critical measure of the financial health of the Quality System. It pinpoints areas where application of the system's prevention- and corrective action-oriented elements can have the greatest monetary impact.

1.8 Product Liability Prevention

The purpose of the Product Liability Prevention Program is to minimize the filing of valid claims for customer or third-party personal injury or for performance failure resulting from inadequacies in product design, manufacturing, installation, or service. With the number of such claims mounting to the millions and the judgments in many cases doing the same, insurance premiums for many companies have become prohibitively large. Some companies even have been forced out of business as a result of these two factors. The prudent enterprise must control many areas to forestall any such litigation.

Obviously, the closer any company comes to designing and manufacturing a product that cannot fail and cannot injure anyone under all conditions of use or reasonable abuse, the safer that company is. Of course, at the same time it must be making a product that will adequately satisfy a perceived need. Although perfect success in such an endeavor is unlikely, the careful examination and improvement of product and process designs (see 2.9 Product Design Review and 4.1 Process Design Review) and the thorough application of 1.12 Corrective Action Program and 1.14 Audit of Procedures, Processes, and Product throughout the operation will materially reduce the risks to which the enterprise would otherwise be exposed.

The product may be a very good one in every technical respect, but it is really only as good as the customer perceives it to be. And as customer perception may be significantly influenced by sales personnel, by public and private statements from management (in union negotiations, for example), and by advertising, it is imperative that these sources of potential misrepresentation be properly controlled. Sales personnel should be made thoroughly aware of the actual capabilities of the product and of the exact limits of what they may promise the customer in product performance and service. They should also know the legal implications of any unauthorized quality, performance, service, or application guarantees they may give the customer. Managers must be made aware that, if they make statements implying that the company does not expend every effort to ensure a risk-free product, the court may draw its own conclusions despite other evidence.

Advertising must be carefully constructed and reviewed so that it tells or implies no more than the truth about the product's capabilities. To give the potential customer the impression that a product will last for ten years when the design life is five, or that it may be used by anyone when only professionals can use it safely, is to expose the manufacturer to large and unnecessary risks. Thus only advertising that tells a valid story should be used.

Product liability laws in several states are in the process of change, but at the time of this writing there is no uniform statute of limitations that operates to limit the manufacturer's eventual risk. For this reason, the business should be prepared to defend itself against failure of any product it has ever produced and to react rapidly and effectively when such a failure occurs. This translates into retaining indefinitely the records that demonstrate concern for and control of the safety and reliability of the product, both through preventive efforts and corrective action. Such records might include those of design and development, purchased material, internal processing, packing, shipping, warehousing, installation, service, field performance, failure, and customer complaint.

Of course, not every piece of data, record, or item of correspondence must be retained. The retention or disposal decision should be based on a realistic evaluation of the classes of documents involved. Certainly, those that any prudent person would keep to support a position should be retained; their unavailability might raise suspicion in the minds of the court. Selective disposal of records from within a class can be even more damaging in court.

The producer must also be prepared to undertake a product recall in the event of a safety or performance problem developing in the field. To do this effectively requires specific knowledge of which items are suspect (see 1.9 Material Tracking), of where they are in the distribution cycle and with customers, of what is to be done with them, and of the personnel and equipment needed to deal with them. Also necessary is documentation of the effort and its results, including selected publication to capitalize on the organization's response for future commercial benefit.

Records retention and product recall are directed toward minimizing the extent of and conducting a successful defense against a product liability claim and are, therefore, of vital concern. Our main interest in this element, however, is toward preventing a valid claim in the first place. Therefore, we must examine the Quality System for activities that contribute toward such prevention, ensure their effective and coordinated application, and supplement them with controls in other related areas.

The elements of the Quality System that have a major role in Product Liability Prevention are the following:

Broadly Applicable Elements

1.14 Audit of Procedures, Processes, and Product—Ensures proper product manufacture, the controlled operation of the Quality System, and that all control documents affirm the manufacturer's commitment to the Quality System.

1.12 Corrective Action Program—Corrects any Quality System failures and documents this effort, resulting in improved control over problem sources.

1.5 Documentation Control—Provides the tie between planning and practice and assures that people have the right documents with which to work.

1.10 Quality Inputs to Advertising—Ensures that no improper or incorrect claims are made in advertising that might mislead the customer or change warranties.

1.4 Quality Manual—Provides the written version of the Quality System, which is the basis for audit and control of the System.

1.1 Quality Policy—States, in the aggregate, the quality, reliability, and safety philosophy of the manufacturer as the basis for establishing the intent of his Quality System.

1.15 Quality Program Effectiveness Review—Documents the manufacturer's commitment to making sure that his Quality System works.

1.2 Quality Standards—Provides the visual evidence of the manufacturer's understanding of his responsibility to customers.

Design and Development Elements

6.3 Design and Analysis of Reliability and Safety Studies—Establishes the manufacturer's program to demonstrate concern for these aspects of cutomer involvement.

6.4 Environmental Impacts—Addresses the question of indirect injury to the customer.

2.13 Preproduction Testing—Ensures that new designs are properly evaluated for safety and reliability and revised when necessary before production begins.

4.1 Process Design Review—Attempts to ensure, before the process design is committed, that the manufacturing processes will not introduce safety- or performance-degrading conditions.

1.13 Product and Process Modification—Provides controls over revisions to assure that no liability risks are introduced by this means.

2.9 Product Design Review—Attempts to assure, before the new product design is released for production, that all pertinent human engineering, safety, manufacturability, and reliability considerations have been properly incorporated.

9.1 User's Safety Warnings—Provides appropriate "use and misuse" safety warning labels, placards, notices, packaging, and installation, operator, and service manual entries.

Production-Related Elements

4.7 Manufacturing Quality Plan—Establishes the production quality controls and the necessary management assurance of their effectiveness.

1.9 Material Tracking—Provides the capability of identifying specific items of product that may be suspect for reliability, safety, or quality reasons.

3.9 Nonconforming Material Disposition—Governs the introduction of known nonconforming material into the product stream, preventing any with potential safety or performance risks from further processing.

4.5 Packaging and Packing Control—Minimizes the possibility that such activities and materials might permit product damage or might introduce unanticipated safety, environmental, or performance risks.

4.6 Production Material Handling and Control—Provides assurance that potential liability risks are not introduced inadvertently during production.

6.7 Quality Level Improvement—Documents the manufacturer's effort to correct product problems and to improve the safety and reliability of products in production.

5.3 Received and Produced Quality Data Reporting—Provides the record of inspections and tests performed to assure conformance to specification of parts, components, subassemblies, assemblies, and final product.

After-Shipment Elements

9.15 Audit of Field Quality Activities—Ensures proper installation and service of the product.

9.7 Drop-Shipped Item Control—Ensures control of items reaching the customer without going through the manufacturer's factory.

8.5 Field Personnel Training—Helps prevent the initiation of customer injury through lack of knowledge on the part of distributors, installers, dealers, or service personnel.

9.12 Field Problem Controls—Provides field failure data and reaction to minimize potential hazards.

9.13 Field Problem Handling—Causes analysis of field failures, producing cause identification and elimination.

9.11 Installation and Service Difficulties—Provides guidance for design correction to improve the effectiveness of both installation and service.

9.3 Product Service—Ensures that field personnel have the necessary information to maintain the product's safety and performance capabilities.

9.8 Renewal Parts Control—Establishes shelf control over service and spare parts.

In addition, numerous other Quality System elements contribute to the overall objective of preventing or minimizing liability. It should be apparent from this discussion that the Quality System, wholeheartedly operated, is management's primary weapon in the product liability "war." Of course, sufficient insurance coverage and capable legal talent must be obtained to provide early warning of changes in legal and regulatory matters and to prevent a failure from becoming a disaster. But the best possible situation ensues when there are no failures. The Quality System provides the only chance of achieving such a goal while remaining in business.

As an aside, it is probably unwise to call this the Product Liability Prevention program in your operation. A plaintiff's attorney might use the title to ridicule your entire approach toward protecting your cutomers. Instead, "Product Safety Assurance" or some other positive customer-related title might profitably be used.

While the above discussion has been directed toward the product manufacturer and his liability problem, most of the principles are adaptable to help merchants, institutions, government agencies, or other organizations subject to liability claims. These groups all share, to a degree, concern for the customer's safety and for their own protection against organizational product liability claims and the imposition of civil and criminal penalties on involved individuals.

1.9 Material Tracking

Configuration control is based on the fundamental concept that the product and its associated documentation must correspond—that a part which is supposed to be made to revision A of a drawing is, in fact, made to revision A. A corollary is that we must know specifically which pieces of final product contained the parts made to revision A, which to revision B, and so on. We also need to know which final product items contained those parts made to revision A that were produced by a process different from that of some other revision-A parts.

The main reason we need to know these particulars is to minimize the magnitude of the problem in case later events (a field failure or a legal or regulatory requirement, for example) force us to be able to identify the specific items involved. Recall campaigns, such as those that regularly plague the automobile industry, are just one of many situations in which the producer obtains enormous benefits from knowing exactly which products contain the item in question. This kind of information is also particularly useful when a product is leased by the customer. To keep the customer satisfied, the producer must provide tailored service, including the addition of options or modifications. So it behooves the producer to know exactly what each piece of leased equipment comprises.

In the practical sense, proper material tracking means that the producer and his suppliers must know which lot of raw material is used for which parts or batches; they must avoid mixing repaired items back into lots of unrepaired items; they must avoid mixing different lots or batches of like items in stockrooms or production containers; they must know from which lot or batch came the items being assembled into which final product; they must minimize deterioration of material in stock; and they must know the source and batch or part application of indirect materials in case contamination of product is discovered later.

Once the product reaches the final customer, the producer's knowledge of the history of replacement parts used, of any service performed, and of modifications made may not be required—but, then again, it may. Each producer needs to examine his own situation as it may be affected by regulation, product liability, and customer reaction—and then decide how far to extend his records.

Techniques used in Material Tracking include:

1. Lot integrity control—Lots are identified upon receipt, and pieces are prevented from straying from their parent lots by accident; if they do so for any reason (such as rework or engineering change), they are formed into a separate lot with an appropriate identification tying them to the parent lot.
2. Processing control—Each item or group of items is uniquely identified (serial or lot numbers with color codes, tags, or log sheets) and has a corresponding permanent record of routine and special processing steps applied to it, including such details as heat-treat furnace temperature, time, and location or rework and repair. For parts made in dies (castings and stampings, for example), it is often desirable to include changeable inserts showing date of manufacture, drawing revision to which made, and cavity or tool number. The record includes the identification of lots of indirect materials used in the processes, such as flux, solder, cutting fluids, and oils, among others.
3. Build control—Assembly and further processing information should show just which items or lots of parts, components, and subassemblies were combined to make the product and what process conditions existed at the time of each step through product completion.
4. Inspection and test—Specific results of these appraisal efforts—including rework, repair, and other nonstandard treatment of the items involved—should be recorded.
5. Field activity and modification control—A similar set of records should be kept of installation or erection processes, service, engineering changes applied in the field, and associated inspections and tests, thus expanding the manufacturer's knowledge into the region of the product in use.

The control programs must also be extended, as necessary, back into suppliers' operations or even beyond. This traceability may need to exist (as once was required in my experience in the nuclear industry) from the zircon sand beds off Australia, through the ship carrying the sand to America, through all fabrication and processing into reactor components, and into lifelong use in the power plant.

When Material Tracking is properly performed, the costs of bulk field problems are minimized and, at any stage of production from suppliers to customer use, material problems can be traced to their sources for ultimate corrective action.

1.10 Quality Inputs to Advertising

The subject of quality is especially important in two types of advertising—that which promotes the sale of a specific product or product line and that which tries to build a favorable image of the producing organization, often called "institutional" advertising. In either case, considerable care must be expended to assure that any quality, reliability, maintainability, or safety claims made or implied can be supported, if challenged. Since a statement on performance or quality can provide an unintended implied warranty, this is one of the major concerns of 1.8 Product Liability Prevention and is covered in that element's write-up.

One of the best ways to ensure that personnel preparing advertising copy do not make unwarranted claims is to provide them with factual, interesting information that will enable them to make valid, informed quality statements. To this end, the Quality System must include a plan to respond to requests for such information with useful specifics and to provide the advertising personnel with details of quality exploits, even though unsolicited. Similarly, sales and marketing personnel should receive these items, as they may find the same information useful, even though the material might not be included in formal advertising.

In specific product advertising, the information provided may include performance or reliability statistics for that product or product line, notable improvements in such statistics, comparisons with competition (as a group or individually identified), or technical endorsements by outside agencies. In support of institutional advertising, the inputs might include the receipt of quality awards or commendations, general quality history (perhaps affected by the application of certain quality-control techniques), reliability or warranty experience, specialized equipment capabilities, new programs that have affected or will affect the customer, and so on.

With such an information program operating, the enterprise will experience a reduction in possibly expensive errors by the advertising group, an improved understanding and acceptance of products by customers, fewer product misapplications by sales personnel, and greater appreciation of the impact of quality in attracting customers.

1.11 Product Certification

Product Certification is viewed by most manufacturers as the formal written guarantee provided to a customer that, based on confidence obtained from the review of all necessary objective evidence, a given piece of product or shipment meets all pertinent requirements. In other circumstances, it may represent a similar commitment by the design and development engineers that their about-to-be-released product design is wholly capable of being profitably manufactured and will meet all published specifications and contractual customer requirements.

Certifying the product may provide the producer with a competitive advantage, often dependent on the form the certification takes and its timeliness and accuracy. Some manufacturers who take measurements on samples of the finished product provide certified frequency histograms or distribution-related statistics for their customers. Others provide notarized test or analysis reports or sworn statements of compliance. The form the certificate takes depends on the customer's needs and is often the result of negotiation, competition, or government regulation. The form may also change with time as the customer's needs change. An example of a simple generalized certification is the color banding of resistors. The purchaser receives an implied certification from the manufacturer that the resistor has a resistance within a stated percentage of the nominal value for that class.

If the customer can make any real use of Product Certification, it must be in direct proportion to the completeness of the information received about the product. For example, if the buyer is concerned with selective fit, the distributions of the mating dimensions will determine how much of such activity may have to be engaged in and what to do to avoid unmatchable sets. If the concern is with chemical or metallurgical mixtures or compounding, the composition and/or purity of the purchased material will affect the customer's own processes.

Although it can be assumed that the customer could perform his own analyses, that assumption may be invalid, or extra cost or delay may be involved. The developed trust between the supplier and the customer resulting from Product Certification of this type can thus be advantageous to both. Some companies, recognizing that they themselves will benefit financially from such a program, have introduced financial incentives to induce their suppliers to provide this kind of certification.

The advantages of this program to the certifier are that (1) practicing the philosophy fully produces a high level of assurance that the customer's requirements are really satisfied, and (2) customer trade relations are enhanced thereby. The customer also benefits from full confidence in the supplier's product. It results in reduced incoming inspection effort and associated delays in placing received material into production, along with fewer surprises from problems missed by incoming inspection.

A common failing in such programs is their deterioration with time so that the certifications no longer reflect the "real world." The certifier should audit the program regularly to make sure that customers are receiving the proper information. A companion program to ensure the continuing validity of these certifications is the random lot inspection/test mentioned in 3.6 Supplier Certification and Objective Evidence of Quality.

In the engineering-type certification many details are involved, particularly in the development of complex or sophisticated products. Such a certification includes objective evidence of the product testing in sufficient detail to permit independent evaluation of the results against published or contractual performance criteria. This evidence, coupled with the identification of potential problem areas and with appropriate levels of review and approval signatures, then provides the basis for determination by the Quality component and Manufacturing whether or not to accept the product for production.

The formal certification statement typically will refer to the product test specifications and performance standards employed, the amount and type of field and/or internal reliability testing performed, the extent of use of standard parts, the completeness of approval or qualification of unique components, the degree of demonstrated manufacturability, the attachment of required relevant data, and the provision of test units for independent evaluation.

Formal Product Certification by the Design Engineering department to its Manufacturing and Quality component customers results in a minimum of potential problems with the product going into production, since the engineers are thereby encouraged to avoid carelessness, oversight, or wishful thinking in their testing and interpretation of test results. Having such a program in place as a counterpart to the participative aspect of 2.9 Product Design Review also tends to reduce delays in acceptance of the Product Certification by the Quality organization and Manufacturing, since they will have taken part in the test planning and will have witnessed its performance.

1.12 Corrective Action Program

Many of the activities of the Quality System result in occasional requirements for corrective action. Inspections and tests identify discrepant material that must be reworked or repaired; process control checks uncover nonconformity-producing machinery or operators needing adjustment or instruction; field failures and customer complaint analyses identify customer-visible unsatisfactory conditions; quality cost analyses pinpoint areas of potential savings; audits reveal errors in documents or failures to meet requirements that call for revision of paper work or practices. In general, this type of correction is expected in every kind of industry and business.

In the Quality System context, however, this type of corrective action is necessary but insufficient. Beyond the correction of the immediate problem or nonconformity there is a need to determine the root causes of the failure and to take the necessary steps to eliminate them. Determination of the root causes of a problem may be as simple as observing and testing. Once I was shown a solder pot that intermittently produced nonconforming work. I noticed that it was directly impinged upon by an air blast from a nearby air-conditioning vent, and a simple deflector in the air stream eliminated the cause of the problem. Unfortunately, other problems are not so readily solved. The use of statistically based experimental design programs will normally resolve the difficult ones; the literature is full of examples. (See 6.8 Statistical Technique Application for further discussion.)

Other problem-solving approaches are mentioned in 6.6 Problem Solving. This element assumes that such approaches have already failed to solve the problem or have been called for by other stimuli, but have not been used.

The Quality System Corrective Action Program, then, is activated in two ways. The first is the result of examining appropriate records and finding failure recurrences after corrective action has been taken. The second is the result of failure to achieve corrective action after it has been called for. In both instances the program involves successively higher levels of management on a strict time schedule until permanent, effective, and economical corrective action is obtained or until the manager with full profit-and-loss responsibility decides that nothing more is to be done.

In situations where numerous corrective actions are called for, it can be helpful to use "real time" management techniques. With the data in a computer, aging and severity ordering of the problems can be automatically shown on an instantaneous basis for top management's attention. Only successful corrective action can remove a problem from the list.

The net effect of this approach is that, although any level of management can take such corrective action, only the top executive can stop it from being taken.

Since an economically justifiable approach can almost always be found to achieve the type of correction described above, the decision not to develop and institute such an approach can be justified only on the basis of resource allocation priorities. This decision can properly be made only by top management.

Corrective action plans must include the participation of suppliers and field personnel in addition to all involved internal personnel. Prompt notification to the instigator of the corrective action taken is a fundamental part of the plan. The necessary analysis of records for recurrent conditions must also be planned, with appropriate program activation standards, and can frequently be accomplished by computer. Ordinarily, these plans are subject to review, approval, and followup by top management (as mentioned in Section 11).

The U.S. Department of Defense Quality and Reliability Assurance Handbook H50, "Evaluation of a Contractor's Quality Program", contains the following criteria for evaluation of a Corrective Action Program:

1. Does the program provide for prompt detection of inferior quality and for correction of its assignable causes?
2. Is adequate action taken to correct the causes of nonconformities in products and facilities? In functions—e.g., design, purchasing, testing?
3. Are analyses made to identify trends towards product deficiencies?
4. Is corrective action taken to arrest unfavorable trends before deficiencies occur?
5. Does corrective action extend to suppliers' products?
6. Is corrective action taken in response to user data?
7. Are data analysis and product examination conducted on scrap or rework to determine extent and causes of nonconformities?
8. When corrections are made, is their effectiveness reviewed and are they monitored later?

In addition, your detection mechanisms must be alert to indications that things are better than expected. The corrective-action response would then be to take steps to incorporate the causes for such improvements into the routine operations of the process involved.

If your overall approach to corrective action passes these criteria with high marks, it will, in the long run, reduce "fire-fighting" and its attendant waste of resources, permit the productive use of manufacturing personnel and facilities now dedicated to producing nonconforming product, and reduce both internal- and external-failure quality costs in dramatic increments.

1.13 Product and Process Modification

Within the experience of product development, production, and usage and service in the field, many instances occur when changes in product design are indicated. The same sort of thing happens to processes; production experience and field quality results often dictate changes in their design as well. The organization may adopt a philosophy of "continuous quality improvement"—recognizing that what is acceptable today will not be adequate in tomorrow's environment.

In addition, other situations may call for design changes—competitive comparisons; the introduction of new design features, materials, or components; or the development of new product, process, or control technologies or manufacturing equipment with unanticipated capabilities. Whatever the impetus for change, the organization must have an effective method for accomplishing it. The Product and Process Modification element provides that method.

Usually, product design changes are the jealously guarded prerogative of the Design Engineering department. Many design engineering departments permit change requests to be initiated by others, but insist on the right of approval or refusal through a formal change system. In some companies, manufacturing or installation engineers may make unilateral changes in product designs to improve manufacturability. It is unheard of, however, for the Service organization to have such authority; yet most of them do it without a second thought to solve some pressing field problem. Of course, in such cases, the change normally affects just one or a few items of product.

Process design changes are similarly expected to be accomplished by the manufacturing and industrial engineers. In practice, changes are also made by operators, setup men, maintenance personnel, line supervisors, and by accident. The foregoing relates to suppliers' processes, as well as to internal ones.

The Quality System requires that, when practicable, all product and process changes be thoroughly investigated in advance (perhaps by special test or by computer simulation), be accomplished under controlled conditions, be properly documented, and have their effects measured and acted on as required. The procedure for this element should contain the formal program for obtaining change, the method of measuring the effect of the change (both immediate and "downstream" in time and space), and the criteria for judging the timeliness, completeness, and effectiveness of the change. The demand for revision timeliness often must be tempered with a recognition of the administrative costs associated with frequent small changes. Regularly spaced change dates, with provision for in-between emergency action, may be overall more economical.

All the changes associated with products and processes will fall under the Documentation Control (1.5) element, which is basic to the concept of configuration management. When a manufacturer must know the "change" status of each piece or group of products in the factory or in the field, the formal Product and Process Modification program is a necessity. Conditions dictating that these details *must* be known include leased product for which the manufacturer has maintenance responsibility, performance contracts, and recall potential associated with product liability possibilities.

With this program, the economic risks of small problems becoming large can be minimized.

1.14 Audit of Procedures, Processes, and Product

System Audit, here called Audit of Procedures, Processes, and Product, is one of the six basic control mechanisms of the Quality System. It is, in fact, a basic control device of any proper management system. The concept was developed many years ago in the financial field to give company management and others a measure of how closely specified accounting procedures were being followed. Quality Control managers, seizing on this concept as an additional check on the validity of final inspection and test, began to "audit" the product after final acceptance. Then external procedure adherence and product conformance audits became commonplace in those industries where the government and its prime contractors were customers. Today, the full application of audit concepts to all aspects of quality in all functional parts of an enterprise is quite commonly encountered.

The basic concept, then, can be simply stated as follows: To identify system weaknesses and improper practices for ultimate correction through the regular comparison of products, processes, and the activities of all personnel involved in the Quality System with preestablished requirements and standards.

The development of an audit plan involves describing the criteria for determining who will audit, the formulation of an audit objective (with associated standards for identifying actions or inactions causing quality problems and for defining the effects that will arouse more than the minimum level of concern), the training of the auditors, the areas to be audited, the scheduling of the audit and notification of personnel to be audited, the establishment of data generation and results reporting, the method for initiating corrective action, and the preparation of any audit checklists. The eight criteria for evaluating a Corrective Action Program, listed in the write-up for element 1.12, are illustrative of what may be included in an audit checklist. An example of an Audit Planning Work Sheet is shown at the end of this element.

A case can be made for using professional auditors to do procedure and process conformance audits for ensuring more consistent results and reduced audit time. An equally strong case can be made, however, for rotating nonprofessional membership on the audit team. The latter alternative serves to educate the auditors, and secondary corrective actions often take place in their own areas because of what they have seen. It also can reveal "pet" problems that might escape the professional auditors, and it encourages acceptance and involvement by the audited areas. These last two results occur particularly when the team includes a key representative of the area being audited. The rotating team concept is even more effective when the chairman is permanent and is also a professional auditor. Product conformance audits are typically better done by professionals.

The multifunctional rotating audit team places a strong continuing requirement for auditor training on the audit program management. As with new professional auditors, each one should complete a formal training program before undertaking the first assigned audit.

Areas to be audited include those found deficient on previous audits, suppliers, internal functional departments, production operations, dealers, service agencies, and so on. An audit schedule to cover all Quality System areas at least once in a given time period, with an entirely different approach used for product audit, should be prepared and announced regularly. The schedule typically covers a year's audit effort and should be announced at least two months prior to the end of the previous year. This gives the people involved a chance to plan for whatever activities and resources the audit will require of them.

As part of preparation for the audit, team members or other planners should prepare checklists, when appropriate, to serve as the minimum definition of the audit scope. Checklists should never serve as "blinders" to the auditors, but can help ensure that key items are not overlooked.

Each auditor should be required to complete detailed audit report forms describing by paragraph number, product characteristic, or other specific identification each condition audited and the exact nonconformance found, if any, with a numerical index rating of importance. The report should also describe any circumstances where the auditor finds the system and its requirements and standards inadequate, incorrect, or nonexistent. The auditor should also state the area responsible, if known or determinable, and any recommendations for corrective action. While the audit emphasis is on finding nonconformances, it is important that the auditors document any evidence of noteworthy commendable Quality System activities—those efforts that go beyond procedural requirements in promoting quality and safety.

Before leaving the audited area, the auditor (or audit team, if working in the same area) should report all findings orally to the audited area management personnel. This report must include any safety practice violations, safety hazards, or environmental problem areas. The individual auditor comments should be compiled by the team into a comprehensive report showing both commendable situations and the specific nonconformances observed, an overall numerical rating or ratings developed from the individual auditor ratings, and a trend of ratings from this and previous audits of the same area. It should also include any team recommendations for corrective action, identification of individuals or functions responsible for initiating indicated corrective action, and a requirement for development of acceptable corrective action plans.

Written audit reports should be issued within two or three days after completion of the audit to permit timely corrective action. Plans for that corrective action should also be submitted by the responsible parties in a reasonable amount of time; typically, ten days is sufficient for corrective action to be completed or, at least, to have the plan for it prepared and approved.

Summary reports to top management of audit trends and corrective action status, highlighting especially commendable positive activities, should be issued regularly. This is often done quarterly.

All audit reports should be placed in the audit files and used for audit planning, trend calculation, and corrective action followup (which may include re-audit), as well as for comparison with discrepancy data from other sources to improve the relevance of the audit program.

When the System Audit is properly used, it provides an ongoing measure of the degree of reality of the Quality System, thus permitting management to make any necessary adjustments to achieve maximum favorable results from the application of the Quality System. It also helps identify areas for improving the process of managing the Quality System.

AUDIT PLANNING WORK SHEET

1. Type of audit (system, procedure, process, product, supplier) ______

   ____

2. Audit scope (SOPs and paragraphs, operator instructions) ______

   ____

3. Checklists and evaluation guidelines required for audit team ______

   ____

4. Required references (previous audit reports, SOPs, OIs, Dwgs.) ______

   ____

5. Requirements for audit team members ____________________
____

6. Training requirements for audit team ____________________
____

7. Scheduled team review date for checklists and criteria __________
8. Scheduled audit date ______________________________
9. Personnel scheduled for entrance interviews ______________
____

10. Notes of audit team meetings during audit ________________
____

11. Corrective action recommendations for closing interview and report
____

12. Notes of closing interview (corrective action schedule) ________
____

13. Scheduled date for audit report distribution ______________
14. Distribution list ___________________________________
____

Note: Items 2, 5, 6, 10, 11, 12, and 14 usually require attachments.

1.15 Quality Program Effectiveness Review

Since the Quality System is designed to be effective and economical in operation, it is important to measure its impact both overall (by the six System control measurements, listed on page 25) and by looking at its programs individually and in related groups. The first step in measuring the effect of anything is to identify what its quantifiable results will be in operation and then to set goals for achieving worthwhile changes in those results.

Some elements of the Quality System cannot logically be assigned quantifiable results; they make the Quality System work or work more effectively, but they have no direct impact on quality, cost, or time. Since these are the areas the Quality System affects and where it is normally measured, those few elements must be viewed as intangible. There may be installation or "housekeeping" ways to track the application of the intangibles, even if their results cannot be measured.

In all other cases, management should establish the specific result areas associated with the elements, usually adopting concrete suggestions from the functions to be measured. Examples might be time from design concept to full production, received material quality index, manufacturing yield, or warranty cost. An efficiency measure might be suitable in some cases; for example, proof of control measures prior to use, percentage of transfer of preproduction quality measurement concepts to production or to other preproduction programs, inspection or test error or bias, current applicability of Quality System documents, or correct use of Quality System documentation. The measurement chosen should always be of genuine interest and utility to the organization. Since making the measurement will cost money, there must be a real benefit both in meeting an improvement goal and in knowing that it has been met or by how far it was missed.

All departments, product or project groups, service agencies, and other appropriate entities, as well as the individuals within them, should establish their operating Quality System goals (with their achievement plans) each year and measure their progress against them. All new products or manufacturing processes should have reliability, yield, and quality-cost goals established for them and the results determined. All processes should have continuous quality-improvement objectives, with specific interim goals. Sometimes it will be useful to establish measurements and goals for partial or complete subsystems (e.g., Purchased Material Control or Customer Contact) to assess the impact of the entire array of related activities. All these goals should be compatible with the overall objectives of the Quality System and with each other; they also should be oriented to ensuring ultimate customer satisfaction with product and service. The method for stating such goals and their derivation from and relationship with the organization's objectives are described in 1.3 Quality Objectives.

One question of importance to virtually all companies is: What impact do quality, reliability, and safety have on sales? Although this is a difficult measurement to make, it is one that should be attempted. Known quality changes can frequently be detected in shifts in sales quantity and patterns, often by date. The correlation between sales trends and the Customer Satisfaction Measurement (9.14) index trends is often statistically significant. This implies a positive relationship between quality and sales, but it is not universally demonstrable. Each company should verify the existence of this relationship for itself. Studies by some market research organizations do not support the contention of a quality-sales relationship; the actual sales patterns should be developed for a given company to reach a conclusive answer.

The measurement of the effectiveness of a program may be accomplished in several ways, either by routine measurement, by special studies including questionnaires, by audit, or by inference from other measurements. It may also be desirable to take several measurements of the effectiveness of a program, perhaps stratified by management level, preproduction or production, customer type or region, or by plant location, among other possibilities.

The purpose of the Quality Program Effectiveness Review is to ensure that all the agreed-upon goals are set, performance measurements are taken as scheduled, comparisons are made, and that action to correct the program or to improve the practice is immediately undertaken when indicated. Such an effort will keep the Quality System operating effectively and will result in the company's achieving its quality and quality-related profitability and competitive position objectives.

1.16 Quality Organization Evaluation

Over the years, particularly following the World War II explosion of quality control applications in industry, there have been many debates on the organizational aspects of the quality control function. At what level and to whom should the Quality organization report? What is the most descriptive title? What are the roles of inspection, test, reliability, service, and salvage? What is Quality Assurance, Product Assurance, Product Integrity, or any other group in relation to the Quality Control component?

There is really only one "right" answer to these questions. The Quality organization should perform those activities that are part of the Quality System and that are more effectively and efficiently performed by that component than by others. In addition, the Quality component must operate the control functions that ensure the continuing validity of all parts of the Quality System. If incoming inspection is assigned to Purchasing, Quality must have a supplier quality audit function; if product inspection and test are assigned to Production, Quality must perform product audits; where life testing is an Engineering function, Quality must perform sample accelerated and normal life tests.

The titles chosen for the Quality components should reflect their actual activities and functions, but the titles should also take into account possible connotations and interpretations within the enterprise and outside it. They must be realistic but reflect understanding of the public relations impact. A man who opened a specialty dress shop with an excellent selection of good merchandise, but called his place "The Fat Ladies' Shop," sold very few dresses. Likewise, your "Flaw Finders Team" will experience little success in obtaining the enthusiastic cooperation of the rest of your staff.

Reporting level(s) and locations must be consistent with accomplishing the company's objectives in the quality area. The managers of the Quality components must be able to perform as the Quality System implies; the reporting pattern must further successful performance, not hamper it.

In all of the other element descriptions there has been no identification of any Quality function by title. This is deliberate. What is required is that each identified job within the Quality System be done effectively and economically—who does it is immaterial. However, the probability that the job *will* be done effectively and economically is often influenced by organizational considerations—thus the need for this element on Quality Organization Evaluation.

Circumstances change. Product liability, government regulation, and competitors' actions, among other influences, provide external pressures on our operations. And our own formal and informal organizations alter with time. Key people come and go; new products are introduced and old ones die; new management concepts are tried; and we grow. What was an optimum Quality component only a few years ago is no longer applicable. We should be alert to the need for constructive adaptation to these altered circumstances.

To be most effective, we should also anticipate new circumstances as much as possible. Therefore, even in the absence of an obviously changed situation, we should make a regular, roughly annual examination of the Quality components—how they relate to all other functional groups in the company and how they respond to the external forces shaping the marketplace. We are concerned with interdepartmental overlaps and conflicts, with authority, responsibility, respect by others, and external acceptance. To this end, we should be prepared to reassign responsibilities, change titles, and alter reporting relationships to achieve the best results from the Quality System. Such a willingness to change for best results implies that, with respect to their Quality System activities, we will examine all functions of the enterprise—not just the Quality component.

All quality-related responsibility assignments, as shown in policy statements, standard operating procedures, job descriptions, department charters, or other company documents, are subject to review and alteration. Regardless of parent function, all programs that have a Quality System impact should be brought within the control activities of the System and then be subjected to the above reevaluation for continuing utility.

Experience indicates that, given the "right" people in key slots, any organizational structure will work. But no structure will work effectively with "wrong" people in key slots. Therefore, a formal structure should protect the organization from mediocrity in management. This concept leads to the "checks and balances" approach to organization and tends to distinguish the larger, established companies from the smaller, more entrepreneurial ones.

The Quality System, with its built-in controls, can be operated successfully in either climate; thus its lack of a doctrinaire approach to organization. The key to its success is to search for the optimum assignment of responsibilities within the unique circumstances of each company. All the benefits of the Quality System are derived from the actions of the people working within it.

1.17 Reference Library

The Reference Library element calls for the establishment and use of a reference and retention source on quality, reliability, and safety techniques. Sometimes this can be accomplished by subscribing to abstracting services in these fields, sometimes the organization has an existing library expandable to cover the areas of Quality System interest, and sometimes the Quality function provides a library resource for itself and for other parts of the organization.

Building the library is rather easy; the difficult part is exposing the organization to its contents on a broad and timely basis, with regular reinforcement and ready retrieval. This is the area where clever use should be made of the latest developments in information transfer. An internal library requires the services of a librarian or an able person to plan and conduct the Reference Library function effectively.

Whatever approach or combination of approaches may be adopted in a given situation, the library will ordinarily not be successful if it is operated as an "on demand" accessory. Circulation of title lists, abstracts, or complete items to involved parties is necessary to ensure some degree of resource use. Sometimes, thoughtful comment on the items may be required of the recipient. What is involved is the recognition that operational benefit can be derived from utilizing the efforts of people outside the organization, as reported in books and magazines. Among the many subjects of interest are new ideas or advances in Quality System management considerations, system engineering concepts, statistical and other reliability and quality control techniques, safety concepts and equipment, government programs, internal and external case studies and success stories, and inspection, test, and control equipment.

Subsystem 2

Product Development Control

It is not the function of the Quality System to specify or to control all of the activities of the Marketing, Research, Design, and Development organizations. But because product safety, reliability, manufacturability, serviceability, and ultimate customer satisfaction with the product are limited by its design, many of these activities must be coordinated with those of other functions and must then be carefully controlled. Product Development Control is oriented to ensure that a new or revised product will satisfy the customer in every respect, and will satisfy the enterprise with its profitability.

From an overall management viewpoint, the elements of this subsystem plus 1.2 Quality Standards, 1.11 Product Certification, 1.13 Product and Process Modification, 3.1 Quality Information Package for Suppliers, and 6.1 Classification of Characteristics, all described in other subsystems, are the minimum list of the vital overlapping concerns of the quality and engineering systems. Other elements call for positive participation by and reaction from the engineering community, which can be seen to have a significant role in the successful operation of the Quality System.

The sixteen elements of this subsystem are arranged in the order of their first appearance in the new product introduction cycle for many manufacturers. That order should not be viewed as rigid, and the elements should provide for their later reinsertion into the flow of activities, as appropriate. In addition, some of them may need to be done only once, perhaps subject to an annual review (for example, 2.8 Tolerance Interpretation).

2.1 New Product Introduction Activity

The preparation of, agreement on, and management according to a coordinated detailed schedule are covered in the New Product Introduction Activity element. The schedule contains all preparation activities of hardware, software, or service design; manufacturing support; materials management; quality planning; field engineering; and others involved in New Product Introduction Activity, including major redesigns of existing products. In most organizations new product design and development efforts are reasonably well scheduled, thus giving the product engineering and marketing groups a basis for planning. Often, however, the purchasing, manufacturing, quality function, and service planning actions are reactive to accomplished designs and occur too late in the development process to be economical or effective. This produces many problems, much unnecessary cost, and sometimes ultimate failure of the organization to introduce successfully an otherwise viable product.

The first thing that should happen, whether the impetus for the new product or service comes from marketing or engineering, is to identify the potential customer and his desires and expectations. This will initiate the first draft of the Product Description (2.2) and begin the continuing focus on the customer, which is critical to the product's success in the market. It permits the designer to select and incorporate new or changed features, as well as considerations of reliability and other product characteristics in use.

Product Design Review (2.9) and Process Design Review (4.1) are oriented to improving the transfer of information from the designer to the other organizations. They also provide useful contributions to the concept of the "robust" design—see the description of "elegance" of design in 2.6 Design Responsibility for Manufacturability. By themselves, however, they do not ensure the recognition of required activities or their proper timing. New Product Introduction Activity is the careful meshing of all these key actions into a single time scale. This permits a proper evaluation of proposals to alter product introduction dates, including recognizing—and planning to compensate for—the effects of bypassing or delaying certain actions.

In developing the schedule, it is important to identify every necessary action of every group involved in or affected by the introduction of the new product. This is a monumental list. It includes, for example, all the product and process development effort; the procurement of facilities, tools, equipment, and material (including spare and replacement parts); the hiring and training of personnel; the preparation of software documentation, instructions, manuals, and advertising material; the design of packages and packaging; the preparation of merchandising, warehousing, and distribution plans; the selection and development of inspection and test equipment and methods; and all the control milestones or actions, such as the design reviews referred to above.

The specifics and timing of these numerous activities will often be different, even for similar products. Therefore, a unique plan must be developed for each product introduction. This can usually be accomplished merely by juggling selected actions within a comprehensive ideal schedule that covers everything and sets a noncrisis but demanding pace for the work.

Among devices developed and used successfully to aid in schedule preparation and management are PERT, PERT/COST, and Critical Path Method. Computer simulation of alternative approaches can contribute to the selection of the optimum schedule that balances risks against time and resources. Many activities, including development of alternate designs, can and should be carried on in parallel with others. Usually this involves some triggering result of development without which the parallel action cannot be initiated; an example would be a procurement action requiring a design decision or a drawing to enable the supplier to begin work.

It is sometimes necessary to delay the initiation of action beyond the trigger point (because of changes in resource availability, for example). When this happens, it means that adjustments must be made to compensate, such as working overtime to hold milestone and end dates. A similar effect may result when there are several necessary triggers; an example might be the start of construction of a nuclear power plant requiring releases from design, from the nuclear regulatory agencies, from the environmental regulatory agencies, from the courts, and so on.

All of these or similar considerations must be factored into the scheduling process, making it very complex. If the product is actually a system comprising several products (for example, software and hardware), the scheduling of interface planning and demonstration activities plus system installation, test, and operation demonstration must also be provided. Plans must further include obtaining internal approval of the schedule, perhaps also from customers or the government, and of the product (e.g., from Underwriters Laboratories as well as from internal appraisal organizations).

2.1 NEW PRODUCT INTRODUCTION ACTIVITY

The New Product Introduction Activity reduces the possibility of an expensive failure to provide for some vital action and ensures the best use of the time between product concept and production. It also minimizes the time for complete step-by-step product introduction and provides the basis for sound trade-off decisions for management seeking to reduce the time still further.

2.2 Product Description

Bringing a new hardware or software product or service to market involves knowing what the customer needs and expects and being able to articulate all those aspects of that product or service that will influence the customer to buy, satisfy all his related requirements after purchase, and make a sufficient profit for the producer. For any complex item, this is a difficult task. Unfortunately, it cannot be done just once per product and then forgotten. As product development proceeds, these characteristics will necessarily be restated.

In addition, the development effort normally takes a considerable time. In the meantime, Marketing often identifies new or modified features needed to satisfy customer demands or to meet competition. Unanticipated applications occur as well, and these applications, plus the application of 2.9 Product Design Review, will result in design modification.

It is important that the company deliberately obtain customer input to this process on a continuing basis, making use of every potentially valid technique to do so. With this information the company can prioritize these features and performance requirements so that the designer may properly make whatever trade-off decisions are necessary. After product development is complete, process development and production may require product redesign, and field experience can result in further change. The design of a product *evolves*—it does not spring "full blown" from the mind of the designer.

This iterative process must be planned for and handled in the most timely and least disruptive way practicable. Of course, at each stage the changes should be made as intelligently and imaginatively as possible to minimize costs, delays, quality and reliability problems, and the need to make further changes. By this means the product may withstand obsolescence for a longer time, thus improving aggregate profit and recognition. The Volkswagen "Beetle" was an example of the realization of this potential.

As stated above, the first stage in defining a product involves a sensing of what the potential customer needs or might need, coupled with a technical breakthrough in research or a "new idea." Either part of this couple may occur first and initiate the other, and they may contribute to the ultimate product definition in different amounts; but any new product other than a copy of a competitor's offering typically involves both of these activities.

"What the customer needs" may be specified by contract. If not, it must be otherwise objectively determined. It is extremely rare that some engineer's or researcher's idea, translated into a wholly new product serving a wholly new function, will sweep the marketplace, unless there is a desire for such a function at least latent in the market. For this reason we must identify all the functions the product must perform, over what time period satisfactory performance must continue, and the circumstances under which this performance must occur. This statement includes what is usually referred to as "reliability"—for which goals must be set (see 1.3 Quality Objectives).

Marketing must identify these functions and the operating and environmental conditions of both normal and abnormal use, and even reasonable abuse, to which the product will be exposed. Then it must communicate these requirements, with finite limits, to the designer. Marketing must also describe the market to be served and obtain competitive product and quality history for thorough comparison and addition of results to the list (see 2.3 Competitive Product Evaluation).

The designer must extend this list to cover the interrelationships of various portions of the design with each other and with the processes that will be used to produce and to install and service them. He will also expose the design at various stages to scrutiny by others (see 2.9 Product Design Review) who will add to and modify this list. The assumptions on which the items on the list are based should be stated and clarified—particularly, in the definition phase for software.

The list ultimately is translated into specific requirements in drawings, specifications, standards, customer performance guarantees, and other mechanisms for communicating these design requirements clearly to those responsible for producing, inspecting, testing, installing, and servicing the product and its component parts. Recognizing the usual division of responsibility existing nowadays between Design and Production, the output of Design is limited to a statement of the results required and the quality, reliability, safety, service, and "cost of use" goals to be achieved. How these results are obtained (the manufacturing, suppliers', and service processes) and how they are proven to exist (inspection and test) are left to the appropriate supplier, manufacturing, quality, and service planning groups to define. The designer's responsibility, requiring close coordination with these and other groups, includes ensuring that every aspect of his product is manufacturable, capable of being inspected and tested, usable by the customer, reliable, and installable and serviceable—with economy and within the known state of the art.

It is implicit that this activity must be completed in a timely manner. All too often, design documents are not completed sufficiently in advance of production. To have the Product Description, in all its ramifications, issued to schedule is to minimize quality and manufacturing problems and costs. Rapid revision of these documents as experience and competition dictate results in similar benefits and in initial and continuing marketplace satisfaction with the product.

2.3 Competitive Product Evaluation

In 2.2 Product Description, a charge is placed upon the marketing function to obtain competitive product and quality history to help define the characteristics of a planned new product. Competitive Product Evaluation is the mechanism through which this is accomplished, providing a broadly based ongoing view of competitors' product features and capabilities for sales purposes, as well as a particularized one for use in a specific product design or redesign project.

Useful information about competitive products can be obtained from competitors' advertising, from the experience of common customers, from trade shows, from dissatisfied customers of competitors, and from various publications of government, general business, and the trade. Often, however, such sources may not provide sufficient valid information or trends in enough detail to enable the designer or salesman to use the results without reservation. It is often desirable, then, to obtain unbiased samples of competitors' products and to subject them to exhaustive evaluation and comparative testing, for reliability, safety, and serviceability, as well as for performance, to provide the basis for realistic product design and redesign and sales promotion decisions.

Although such test results often must be carefully restricted in distribution, particularly if they are unfavorable to the tester or could jeopardize the tester's company in its public or regulatory dealings, they must be used by management at least to initiate appropriate research and the resulting design changes sufficient to cope with the identified competitive threat. Some companies do no innovative research and design work at all, choosing to concentrate their design efforts in reacting rapidly to new ideas surfacing in the products of their competitors with similar offerings of their own. Since they avoid a significant portion of the innovators' large initial investment and continuing expense in research and development facilities and personnel, they can be inherently more profitable at producing long-running products than their industry's technical leaders.

Such a strategy forces the enterprise adopting it to concentrate strongly on this element of the Quality System. For greatest impact on profitability, the strategy should also include heavy emphasis on high productivity and quality manufacturing capability—with the capacity for rapid change to new product configurations. The manufacturing capability portion of the strategy would be, of course, as beneficial to the technical leader as to the follower.

Even companies with a virtual monopoly in their product areas need to keep an eye on competition. An organized program to do so, as described above, will minimize the market impact of product surprises by competitors and prevent the user from indulging in self-destructive complacency.

2.4 Reliability Planning and Incorporation

One of the subjects included in the term "Quality" as it is used in this book is that of Reliability. As a description rather than a definition, Reliability can be said to be "Quality over time." In other words, a reliable product is one that continues to produce satisfactory results for the customer throughout its intended life cycle—in the customer's environment and under conditions of use or "reasonable" abuse.

The hardware, software, or service product designer has the responsibility for determining, with help from other internal functions, suppliers, and even customers, the real reliability needs and expectations of the customer—including any contractual specifications or internal management goals for reliability, the environment within which the product will operate, and other conditions of use or abuse that the product may encounter in service. The designer must then take steps to include testable (see 6.3 Design of Reliability and Safety Studies) characteristics in the design so that the reliability can be simulated or demonstrated as a prerequisite for design release for production.

Techniques that have been developed to assist the designer in satisfying all these requirements include those of Redundancy, Operating Condition Derating, Failure Mode and Effect (and Criticality) Analysis, Step-stress Testing, Accelerated Life Testing, Fault Tree Analysis, and Block Diagram Analysis, among others. An excellent reference for these and other reliability design matters is *Practical Reliability Engineering* by Patrick D. O'Connor (New York: John Wiley and Sons [a Wiley Heyden Publication], 1983).

The Reliability that is designed in is, in virtually every case, the limit that can be achieved. Manufacturing a hardware product and, to a degree, software, as well as providing a service, is typically doomed to produce something of lesser reliability than that limit. That is why element 2.6 Design Responsibility for Manufacturability is so important. The design decisions that are made all have the potential for "elegance," including meeting the reliability requirements. (See 2.9 Product Design Review.)

An organization that appropriately addresses Reliability Planning and Incorporation as a key part of the Quality System can expect to achieve significant levels of customer satisfaction with both the product's performance and the long-term costs of ownership. Of course, avoidance of legal actions based on "failure to perform" principles is also a desirable outcome.

2.5 Safety Engineering

In a manner similar to that of 2.4 Reliability Planning and Incorporation, considerations of safety and security in production, handling, storage, and eventual use of a product must begin with the design. The designer must take into account all likely alternatives of use, and even abuse, of the product and its components—not only when it is finished and in the customer's hands, but while it and its components are being worked on or transported by a supplier, the Production people, a distributor, or a dealer.

For software products, protection must be provided to ensure that damage to, pilferage from, or unauthorized modification of the software and its files are not possible. If this cannot be assured, then the program must at least incorporate alarms that will indicate the source of the problem.

During product design and development, planning must begin for the manufacture and assembly of the components, the eventual product production, distribution, and service, and associated personnel safety provisions. In addition, as planning proceeds—and, often, as a general matter not limited to the introduction of a new product—operational, service, and user-safety considerations must be addressed.

The federal Occupational Safety and Health Administration (OSHA) has issued regulations covering employee exposure to potentially hazardous materials and practices in industries of many types. As might be expected, materials that may be hazardous to the health of any employees of a company using or producing them are also potential pollutants of the environment. It is therefore impossible to separate this aspect of safety from environmental concerns. Anyone working within this element, then, should be attentive to the considerations of 6.4 Environmental Impacts.

Whoever is responsible for Safety Engineering, in addition to being aware of the environmental effects of the processing materials and the effects of the environment on the processes, must be especially sensitive to the mechanical safety questions of day-to-day existence in the plant and office operations for which he is responsible. In this aspect of his role, he must be concerned with machine guards, protective clothing, breathing apparatus, equipment interlocks, explosion-proof electrical fixtures, deburring processes, safe handling devices, and myriad other situations that can trap the unaware or careless employee. He must also know the proven mechanisms for preventing the associated possible injury or death. The tie with 4.12 Housekeeping and Plant Safety should be obvious.

One aspect of the job that is often overlooked is the necessity for new or transferred employee safety orientation, training, and retraining; continuing emphasis on safety with and by supervision in all areas; and incorporation of safety material in the coverage of 8.13 Quality Information Transfer. Safety consciousness throughout any organization pays off not only in reduction of the time lost from industrial accidents but in useful caution being exercised by employees away from the job. It also helps engender the prevention orientation so critical for everyone involved in the Quality System effort.

2.6 Design Responsibility for Manufacturability

In many large companies, and even in some small ones, there is a tendency for product designers to concern themselves solely with the technical excellence of the design. They often ignore or subordinate the capability of purchasing to procure parts of requisite quality at a fair price and of Production to produce the new item at reasonable total cost, the ability of inspection and test activities to be performed economically, and the serviceability of the product in the field.

The product development process is one of repetitive sequences of trials, obstacles encountered, technical problems defined, solutions developed, and problems solved. In this process, trade-offs are often made, sometimes in the interest of increased manufacturability, but more often to permit the use of a nominally less expensive component.

Experience shows that virtually all technical problems encountered in the development process have more than one possible solution. The same experience shows that at least one of these solutions is "elegant"—that is, it will not only solve the technical problem but will enhance manufacturability and serviceability with equal or better reliability, safety, and production costs. The development engineer, however, rarely searches for an "elegant" solution because he is not held responsible for doing so.

Instead, the engineer is universally held to a schedule, frequently set in response to a real or suspected market situation, and is usually held to a cost goal for direct material and direct labor. Almost always, the schedule covers the period from design project initiation to "release" for production; that is, the engineer's schedule-meeting performance is measured by the ability to turn over to Production the necessary drawings and specifications on a certain date. Instead, this aspect of performance should be measured against the date when scheduled full, nonconformance-free production was obtained.

As to the cost measurement, the engineer should be measured against appraisal and internal and external failure cost goals in addition to direct material and direct labor. With these other costs (from 1.7 Quality Cost Program) included, the engineer would find it advantageous to find the "elegant" solution to each technical problem.

In those cases where the designer has failed to develop the "elegant" solution, the principles of "Value Engineering" or "Value Analysis" should be applied. Simply stated, these principles require that the function of the item (part, subassembly, assembly, or end product) be stated in the customer, product structure and performance, and manufacturing contexts. Then it is necessary to find the most cost-effective way of performing that function.

Recently, a similar technique was developed in Japan and introduced in the United States under the title of "Quality Function Deployment," or QFD. QFD places more emphasis on the customer's considerations and provides a mechanical method for organizing the decision processes, but the ultimate intent is the same as that of "Value Engineering."

To accomplish these results, the designer must assure that, at every stage, the product is defined so that each of its characteristics can be produced and the production results evaluated, both at optimum cost. In cases where production of the characteristics involves processes or controls not within the experience of the producer organization, the designer instigates appropriate development effort, including qualification testing, to ensure that the capability will exist in time for scheduled use. In addition, where the product and process are interdependent, such as in the production of bulk chemicals, the designer will ensure that the producer fully understands the input-output relationships between them. This can often be explored by computer-simulation techniques.

When the design is "elegant," inspection and test costs are minimized, as are production problems and associated costs, and the time from the new-product concept stage to full production is reduced. An example of how this concept was effectively applied is provided in the descriptions of 2.9 Product Design Review and 4.1 Process Design Review.

2.7 Tolerance Partitioning

When an engineer produces a design, certain key aspects of the product are identified. Such items might include physical dimensions, component values, finishes, or virtually any measurable condition that could materially affect performance, safety, or saleability. Other design details that the designer views as not critical are left to the draftsman to define.

The designer or draftsman first establishes a basic or "nominal" condition for each characteristic. Properly generated, the nominal values are those which, if every characteristic were at nominal, would cause the product to exhibit perfection in all aspects of production and customer experience. This is the basis for the concept of continuous quality improvement: to reach the nominal value of every characteristic, with zero variability. Unfortunately, we cannot expect to achieve that condition immediately upon the start of production, so the designer or draftsman must also establish the tolerance band within which the manufacturer is required to produce it. The engineer typically makes tolerance decisions in only a very few critical cases; the draftsman determines the great bulk of them.

Tolerance partitioning is often done through reference to drafting standards, set by the organization, which do not reflect genuine manufacturing capability in any case but are often a reasonable approximation of it. Unfortunately, the exceptions are numerous and the net effect is to penalize production and overall cost effectiveness in the interest of simplifying drafting effort. Sometimes, when there are no standards, the draftsman relies on past practice for similar items to guide decision making—again, often without relating it to demonstrated manufacturing capability.

The problem becomes particularly acute when a series of dimensions has an additive effect, a "stack-up" condition, which ultimately must meet a limiting requirement. A conservative design approach, "worst case" tolerancing, has been used at times, with the resulting performance or assembly-limiting tolerance divided among the contributing units so that the interaction or combination of items at the extremes of their tolerances would not exceed the stack limit. When there are just a few contributors, this approach occasionally can be realistic; when there are many, or the stack tolerance is small, the result is highly uneconomical. This is the easiest technique for the designer to use, however, and if, despite the relatively high cost often associated with them, there are manufacturing processes capable of producing to the resulting tight tolerances, the designer can feel very safe with the design from a product performance viewpoint.

The frequent specification of literally impossible-to-meet tolerances resulting from the "worst case" approach has caused designers to search for satisfactory alternatives. One technique commonly adopted has been labeled "statistical tolerancing" and it takes two forms. In the first, which is rather easy to apply, the designer assumes that processes will be available for all independently contributing dimensions and that they will produce normal distributions of results with averages at the centers of the tolerance bands and for which the $\pm 3\sigma$ bands coincide with the tolerance limits. The designer then can use the additive properties of the variances involved, according to the formula $\sigma_T^2 = \sigma_1^2 + \sigma_2^2 + \sigma_3^2 \ldots + \sigma_n^2$, to assign values to the component variances that will produce a 6σ spread corresponding to the stack tolerance limits. Given some flexibility in making these assignments, the designer can compensate for known process problems by adjusting other variances, as long as the total is unaffected.

In the second form, the designer deliberately obtains, from existing data or through special studies, statistical measures of each of the involved processes and determines the probable criteria affecting their combination and reflecting the realities of the processes. One example could be from a highly skewed distribution with an average far from the midpoint of the tolerance band, although its range might correspond to the tolerance limits. Another example might involve nonindependent processes in which a "high-side" dimension on one part will force a balancing "low-side" dimension on a mating part. Such a situation often happens in plastic molding operations.

Such "realities" could cause the designer to adopt statistically valid approaches other than the central limit theorem application implicit in the above-described version of statistical tolerancing. Techniques available include those covered by David H. Evans in the three-part article, "Statistical Tolerancing, State of the Art," in the *Journal of Quality Technology* (October 1974 and January and April 1975).

As these tolerance assignments are made, the designer will of course include considerations of economy throughout, not imposing unnecessary requirements anywhere but making practicable adjustments to promote fault-free manufacturing, servicing, inspection, and test of the product.

The net effect of this program will be to provide, for any design, the lowest potential cost and nonconformance level in production. Of course, as the organization comes closer to achieving the objectives of continuous quality improvement, the above considerations diminish in utility.

2.8 Tolerance Interpretation

Historically, drawings and specifications of one type or another have been used by designers to inform producers and quality control personnel of the product's needs. These documents have satisfied this intent with varying typical degrees of success, rather poorly when the products are complex but perhaps somewhat better with simple items. The manufacturing and quality planning people, including supplier, internal, and field personnel, have responded to this lack of complete information by asking the designer to provide more information (through 1.2 Quality Standards, 2.9 Product Design Review, or 6.1 Classification of Characteristics, for example) or by making their own judgments.

In "own judgment" decisions, the opportunity for catastrophic error exists, since no one can be expected to be as knowledgeable about the nuances of the design as its designer. Therefore, within the Quality System the orientation is toward the first alternative. To accomplish this often requires the cooperative development of the required information, as the designer is generally unaware of the unspecified conditions needing clarification.

A common extension of the drawings and specifications is what is called "Workmanship Standards," sometimes developed by and for an entire industry. This is described under 1.2 Quality Standards. Another portion of the needed information is covered by this element and is often referred to as "shop run" tolerances. Here the manufacturing planners, the quality function planners, and the designers define jointly what is meant by those conditions that must occur on the product but do not show up in the drawing or specification because of design convention or oversight.

An example might be a simple turned shaft with a hole drilled through it perpendicular to the long axis. A common drafting convention would show the centerline of the hole intersecting the centerline of the shaft with no displacement tolerance specified. In practice, both the machine shop personnel and the inspector need to know how much displacement is permissible, and the designer should be concerned with how much displacement is likely. The mechanics of the tooling, fixturing, and prior operations and the effects of materials and operators all contribute to the latter. By reducing this and similar situations to an empirically determined but technically satisfactory standard, all parties can have their information needs satisfied.

Another aspect of Tolerance Interpretation involves both published and local standards. The literature addresses the question of measurement sensitivity by defining two available interpretations of excursions beyond the tolerance limits, involving only digits further from the decimal point than those specified—for example, a measurement value of 4.245 against a tolerance of 4.24–4.20. The two standard interpretations are the "absolute" (the example would be out of specification because an infinite number of zeros is understood to the right of the last significant figures) and "rounding off" (the example would be in specification because a 5 is rounded off to the nearest *even* significant digit, in this case to 4.24). An organization should decide on one of the two interpretations and apply it everywhere: in design calculations, in manufacturing planning, and in inspection and test. Administratively, the absolute interpretation is simpler and much more clean-cut, minimizing internal problems and those with suppliers, service agencies, and customers. Rounding off *appears* to grant the producer more leeway, but if the designer is planning in this framework, in practice it frequently results in a net tightening of the tolerance.

Rather than use either of the two standard approaches, however, many companies and even individuals apply their own local interpretations. I have known inspection personnel, including managers, who apply from 10 to 50 percent of the tolerance spread to each limit of the tolerance as an unofficial allowance. The explanations for this practice vary from "engineering always overspecifies" to "the material review board always accepts to this limit; I'm just saving them time and effort."

The important thing about all of these approaches is that everyone involved should understand and follow the same rules, as they are all workable if consistently applied by everyone. The unofficial decision rules cause the greatest difficulty precisely because they are usually unknown to the greatest number of involved personnel.

Once an organization selects, disseminates, and enforces a single Tolerance Interpretation, with a thorough statement of shop-run standards, a significant reduction in the number and severity of unexpected quality problems will develop. With all planning and operating personnel "reading from the same sheet of music," the likelihood of harmful misunderstandings, with associated costs, is significantly diminished.

2.9 Product Design Review

The foremost consideration in the mind of every designer is that the product being designed must have technical excellence; it must do what the designer intends it to do. A close second is cost; the labor and materials required for the product must not exceed some limit, perhaps not well defined or inclusive of all pertinent costs. Other considerations—manufacturing yields, inspection and test capability, shelf life, reliability, safety, serviceability, cost of use or ownership, human engineering, patent position, environmental impact, conditions of misuse or special use—have varying impact on design decisions, as a result of the type of product, the market, and the expertise of the designer and the supervisor. Unfortunately, in many cases some of these lesser considerations, although vital or at least important, may be completely overlooked or discounted by the designer.

For this reason, and to ensure that proper attention is paid to all of these considerations, some companies have instituted programs for requiring their inclusion in the design. These programs have taken several forms, but they all fall under the concept of Product Design Review. In the older type of quality control program, this typically meant that the design had to be reviewed and approved by the quality control manager and staff before release for production. Design engineers usually resisted and often refused to participate in such activities because they could not accept the idea of "superiority" resting in the Quality Control department.

The Quality System, not concerned with generating interdepartmental feuds, incorporates the idea of having such reviews involve all interested agencies—Marketing, Packaging Design, Purchasing (including suppliers, where appropriate), Manufacturing Technical Support, Production, Safety, Reliability, the Quality function, Installation, Service, other Design personnel, customer representatives (if this input is not provided by Marketing or Service), Legal, and any others. Experience indicates that the best results are obtained by having reviews conducted, perhaps with increasingly broad participation by the above groups, at milestone points throughout the design and development cycle.

The first review of any type to be held for a new product is frequently of the type that is often called a new product or technical review. Here new ideas are presented to top management for their consideration and direction. This is often done prematurely and may result in direction being given for precipitous action with undesirable cost and other consequences. It is important, however, to make major new-product concepts visible to top management; it never should be done before the design engineering community is prepared to support the project properly.

The first Product Design Review would occur at the conceptual stage of design and involve relatively few of the different disciplines and, if practicable, should occur before the first technical review. The last review of many would involve all functions and would occur sometime after the start of production. It would include examination of field results and the accomplishment level of all goals established for the product.

Design reviews should be quite detailed and specific. To permit this to be accomplished efficiently, the functions participating in the review should have sufficient time to evaluate the design and to prepare their comments and questions in advance of the actual meeting. One useful approach has been to hold an initial meeting where the designer(s) explained the design to representatives of all involved functions and gave them copies of drawings, specifications, or whatever design documents existed at that stage. Two weeks later the functional representatives present the results of their functions' evaluations of the design to the designer(s) in a similar meeting, emphasizing the specific areas of their functional responsibility, but not necessarily exclusively. Group discussion then provides firm and specific direction to the designer, who is bound by such direction unless it can be proved to the review chairman that the decision of the review team was technically wrong or not economically justifiable. To minimize potential charges of protectionism, the review chairman is typically the Quality function manager or some neutral person of substantial rank.

Starting in 1965, I was for several years in charge simultaneously of design engineering, manufacturing engineering, and quality control for a manufacturer of sophisticated electromechanical devices. The new-product development effort was organized into four-man teams, physically stationed together, each team made up of a design engineer, a manufacturing engineer, a quality engineer, and a development technician. The product, the processes, and the inspection and test programs were developed concurrently and interdependently. The results were a product that satisfied all technical requirements, total costs no higher than two-thirds of initial approved estimates, elimination of human inspection and test except limited patrol inspection and final lot acceptance, no lot failures in ten years, and complete customer satisfaction with the product.

One significant feature of the program was that every development decision involving a potential design change in either product or process was subjected to design review. The design engineer stated the problem and proposed a technically oriented solution. The team then searched for an alternative solution that would satisfy the technical question but would, unlike the engineer's solution, also tend to improve the fault-tolerance of the design and thus its manufacturability, reduce cost, and increase reliability at the same time. This reflects the fact that there is always more than one solution to a technical problem and that of the alternatives there is usually at least one that is truly "elegant," satisfying all the above criteria. Interestingly enough, an elegant solution was always found, subject to the design engineer's demonstration; after both the designer's solution and the selected alternative were tried, the alternative solution was, indeed, seen to be elegant. By this process, products having all the characteristics cited above were developed on schedule.

This involvement of personnel other than product engineers in a design team is a key feature of effective design review. In recent years, the concept has been called "simultaneous engineering," "co-location," and the "design team approach." These special-interest personnel need not necessarily be full-time on the project, but they must be regularly and frequently involved. In complex product situations, supplier, service, and even customer personnel may be involved on the basis of at least a day's time every week or two—in addition to the obvious participants representing Quality and Manufacturing.

With all of the concern expressed so far for ensuring that these other considerations are incorporated by the designer, it is important to include in this element an effective mechanism to make certain that identified changes are made expeditiously to the design and its documentation. In addition, where appropriate, the design should be oriented to automated or at least mechanized manufacture, handling, inspection, and test.

Part of the design review activity involves recalculation of the key decision processes used by the designer. In certain cases this will require actual experimentation to establish the validity of assumptions, particularly regarding those product characteristics that are process-dependent. This interrelationship between the product and the process, particularly of importance in chemicals and pharmaceuticals, is an area requiring careful examination and cause-and-effect determination. Even in mechanical product areas, "selective fit" conditions fall in this category, since they are expensive to operate under. The companion activity of 4.1 Process Design Review should be carefully coordinated with the work of this element.

Product Design Review is an important contributor to the evolution of the Product Description (2.2) and has the ability to reduce costs; improve quality, reliability, and safety, thus reducing product liability potential; and reduce the total time required for the New Product Introduction Activity (2.1) to result in a product achieving forecast yields, reliability goals, and production costs.

2.10 Customer-Centered Quality Audit

One of the six basic control mechanisms of the Quality System is Customer-Centered Quality Audit, a special form of product evaluation because it is conducted from the viewpoint of a typical customer, rather than just through comparison with drawings and specifications. It goes beyond determining conformance to requirements and asks whether the product or service satisfies the real needs of the customer, while giving proper performance, reliability, and safety protection even when the product is misused.

We are concerned with both design and production questions in Customer-Centered Quality Audit, although the bulk of changes resulting from these audits are of a design nature. Desirable, even necessary, features are often overlooked by the product planning or design functions or deliberately omitted through misjudgment. Often, the design decisions do not consider customer facility.

An example is the case of the manufacturer of a two-door automobile who placed the seat back release beside and near the front of the seat cushion. To place something in the back seat demanded two-handed release and tilting back of the seat, posing a hard-to-resolve dilemma for anyone with an armload of groceries in bad weather. To complicate the matter still further, Manufacturing elected to simplify their logistics problems by procuring and installing "left" front seats on both sides, since the only design difference between the two seats was the side on which the back release was located. The result was that exceptional gymnastic ability was required of anyone attempting to load the back from the passenger's side.

The point is that both of these situations met specification under the company's operating rules and would never have been identified by conformance inspection, test, or audit. Only when we look for customer satisfaction, rather than just conformance to specifications, can we expect to prevent such errors from reaching the customer.

Several elements of the Quality System provide opportunities to identify this type of problem; 2.9 Product Design Review, 9.12 Field Problem Controls, and 9.14 Customer Satisfaction Measurement are the most likely. But none of them institutes a specific discipline to accomplish such identification; therefore, Customer-Centered Quality Audit has been established to make sure that this important examination is conducted.

Many companies actually involve selected customers as individuals or as representatives of a group in Customer-Centered Quality Audit along with their own trained personnel. Often the latter take for granted certain product conditions *because* of their familiarity and expertise, ignoring details the customer may find objectionable. The less sophisticated user frequently identifies these things that the technician overlooks or dismisses.

When the auditors find a problem, particularly after the product is in production, the corrective-action effort may take a substantial amount of time to complete. The correction may involve redesign, retooling, selection of new suppliers, and other time-consuming actions. If so, it is often necessary to be prepared to handle customer complaints and special service requirements until the basic problem can be corrected. This is the reason that Customer-Centered Quality Audit should be carried out during the design and development cycle, to minimize field problems attendant upon necessary redesign.

Proper analysis of the audit results patterns and trends can pinpoint system problems, training needs, and areas of needed improvement in initial production definition. But, most important, the proper application of the technique ensures a higher level of customer satisfaction with the product.

2.11 Preproduction Quality Planning

Most manufacturing companies do at least a fair job of planning for the inspection and test of items for and in production (see 4.7 Manufacturing Quality Plan); it is very unusual for much planning to be done for similar appraisal activities associated with product under development. In many cases the engineers do not know the degree of nonconformance of parts and components they use to build their experimental or prototype units. Yet they make decisions affecting the direction for the design to take, and they draw conclusions about achieved levels of performance and the proper values for product characteristics based upon their use of these questionable items.

Some companies supply their engineering stock rooms with rejected items from Production, assuming that the cause for rejection will not affect later engineering decisions. Although such an assumption might be correct some of the time, the likelihood that this practice will never cause any errors or problems is vanishingly small.

The planning for appraisal activities associated with preproduction material and product will inevitably be somewhat, if not completely, different from that for Production. Nonetheless, the two programs should be compatible with each other and Preproduction Quality Planning may well provide the basis for the activities and equipment to be incorporated into the Manufacturing Quality Plan (4.7). The extent of the production plan should provide a model for the preproduction plan. Sophisticated planners will often seize the opportunity to test new equipment or programs in the preproduction environment prior to introducing them into production-related operations. The laboratory conditions which are often associated with product development frequently provide a much more informed capability for determining the real utility of innovative quality approaches or equipment.

Delays in development often occur because the engineers find that they had previously erred in some conclusion; more often than not, the error occurred because a nonconforming or unevaluated substitute item was used to build a prototype, the test technique used was invalid, or the test equipment was either incapable of performing the intended test or was out of calibration. The savings in time and money potentially available from avoiding such errors are typically large. Preproduction Quality Planning, properly completed, followed, and with the results documented, can achieve these savings and help to avoid such errors in later development programs.

2.12 Preproduction Material Handling and Control

Although most manufacturers try to do an effective job of managing production inventories and dealing with production material quality problems as discussed under Production Material Handling and Control, very few manage their preproduction material well. Numerous conditions contribute to this: there usually isn't much of it, much of what exists is of an experimental nature destined for an engineering laboratory, most of that will never reach a customer, and so on. Commonly overlooked is that the engineers use this material to run tests and make design studies upon which fundamental and far-reaching product design decisions are made, often testing items made up of substitute materials or components quite different from the official design.

Years ago I worked for a large manufacturer of highly complex capital equipment. During the development cycle, numbers of full-scale test models had to be built, the data generated by testing each one providing the basis for the design changes for the next. At no time during my first year there did the engineers test a model that met the drawing requirements existing for that stage of development. Key components would not be ready on time, so the engineer would assemble an earlier version, or one hastily manufactured from different materials, or something that would perform the needed function, although it would not be even remotely similar to the correct component. Many of these problems, which contributed to numerous erroneous design decisions, were caused by lack of sufficient attention to material control. The situation was so bad that I was driven to protest that "We never make one alike, let alone two!"

Eventually management realized that delaying a test until the correct components were available produced a valid design faster than testing the wrong thing on schedule. Then the proper attention was paid to scheduling the material, performing the necessary inspections and tests, dealing rapidly with nonconforming items, handling everything to minimize the risk of potential damage, and assembling with care—all with sufficient lead times.

Nonconforming material was moved to an enclosed holding area where rapid disposition was made and correction or replacement initiated. Attention-getting devices such as blinking lights or bright-colored placards were used to expedite movement of critical items. We also cleared out all the obsolete material from racks, file cabinets, desks, and closets in the engineering and production departments. Besides reaping a bonanza in scrap sales, we eliminated a major source of out-of-tolerance additions to our test models. This was followed by periodic audits to ensure that the stockpiling of such worse-than-worthless material did not recur.

Ultimately we found that proper and expeditious handling, inspection and test, and control of our preproduction material reduced our inventory of such material, improved our design decision validity, and reduced the overall time from design concept initiation to the start of production.

2.13 Preproduction Testing

Preproduction Testing ensures that Quality System information needs are met by new product test practices. This concept includes testing of components and product in house and in the field with key characteristics at or beyond tolerance limits, under stress conditions reflecting both use and abuse of the item, and using statistical techniques in designing the test program and in analyzing the results. It also includes thorough measurement, testing, and recording of the quality of the parts, components, and subassemblies used to make the prototype and other preproduction products so that we know exactly what is being tested.

All too often, engineering development tests are run on products built from parts of unknown quality, including substitutes for major components, and the resulting design decisions are ultimately proved erroneous. (See 2.11 Preproduction Quality Planning.) The costs in wasted time and money are substantial, far exceeding those that would have been invested in determining just what it is we were testing and then making appropriate adjustments in design conclusions. They also often are conducted using equipment or methods that do not reflect the real requirements of the product or of the customers. (See 2.14 Test Method Control.)

In Preproduction Testing we seek to provide valid and rapid design guidance and verification to the engineers, a measure of the relationship between the product and the processes used to produce it, and a reference base of specific data within which we may be able to find indications of the sources of later unexpected quality problems. The work of this element also supports the programs of 1.8 Product Liability Prevention and 6.3 Design and Analysis of Reliability and Safety Studies. Therefore, an entire structure of needed practices and records must be built to determine just what is required by this element.

The product development process recognizes several stages of progress toward full production. (See 2.1 New Product Introduction Activity.) Terms used include "feasibility study," "prototype," "preproduction," "pilot run," "field test," and "early production," among others. As the product moves through these stages, most organizations realize that the requirements for control and documentation increase; but the Quality System recognizes more extensive requirements, even in the very earliest stages, than most organizations meet. When one considers that the body of design decisions made during product development is the first line of defense a company has against delayed product entry into the market, excessive manufacturing costs, customer dissatisfaction, and product liability judgments, the urgency of making proper decisions becomes apparent.

To develop a thorough Preproduction Testing program, one must examine the key product characteristics, using 6.1 Classification of Characteristics concepts, and select those for which the knowledge derived from testing at or beyond stated tolerance limits would be most useful. Then expected and abusive operating environmental conditions must be incorporated, where appropriate, at such levels as to represent accelerated life conditions. In such testing, every failure of any item under test must be treated as a design deficiency—requiring item redesign to eliminate its cause. The only exception to this would be a demonstrated repeatable failure of the test equipment.

Since it is usually desirable to test minimum numbers of items, to do so at known risks of error, and to obtain the most useful information possible from such testing, the use of statistically based experimental designs becomes very attractive. Less thorough approaches depend for their success on intuition and luck, somewhat undependable bases for action. Of course, when a sufficient body of knowledge exists as a base for sound decisions, computer simulation can often supplement or even replace much hardware testing that would otherwise be required.

The ultimate intent of the design effort, including testing, is to enable us to place a readily manufacturable, completely customer-satisfying, high-quality product into production. To do this we must study the interaction between the product and the processes to be used to produce it to be sure that they are fully compatible and that we understand the input-output relationships involved. So, at an appropriate point in the product evolution cycle, we must introduce tests that will relate product characteristics to associated process capabilities, including those of all relevant suppliers. This program is described in detail in 2.6 Design Responsibility for Manufacturability.

Since customer satisfaction is obviously one of the intended results of the design program, it usually requires enough field testing of the product (often by potential customers) that product performance, utility, and serviceability can be adequately determined. This sometimes takes many months, and the company cannot always afford to wait before starting production. Steps should be taken to minimize the time gap and, at the least, to incorporate field test experiences into a product redesign as rapidly as practicable.

When all of these considerations are factored into a Preproduction Testing program, there will be significant savings in product development time and in both internal and external failure costs.

2.14 Test Method Control

Test Method Control is aimed at ensuring that test equipment used during the new product design and development cycle is capable of reflecting not only engineering considerations but true customer requirements as well. Obviously, we would also expect that the actual testing would utilize the full capabilities of the equipment.

Ultimately, the proof of the design effort comes in customer application. But we usually cannot expose a new product to such use during the development cycle—for security reasons, if for no other. And at the prototype stage, the product is usually not capable of actual customer use, so our own tests must suffice. With all the critical decisions to be made based on our test results, it is obvious that much depends on the validity of these tests. Therefore, we must be sure we are conducting the right tests.

This means our test equipment and methods must be the right ones. If the product will be exposed to nonstandard environments, we must simulate those environments realistically. If the product will see erratic duty cycles, our equipment must be capable of duplicating them. We must be able to extend the test severity and duration beyond the limits of practice. Our test equipment must be able to accept the product's prototype and production configurations, which are often significantly different.

The selected test equipment and methods should be subject to thorough evaluation, as described under 2.9 Product Design Review, to provide advance warning to the production test planners of the types of things they should plan for as well as to minimize errors in preproduction test. Indicated corrections should be performed before testing starts. Calibration, maintenance, failure plans, and safety considerations should be developed and used to ensure maximum likelihood of successful test conduct and usable results.

Test Method Control gives a company an edge in developing a product that is truly responsive to customer needs, without the lost time or wasted money that results when the product is not developed with such orientation.

2.15 Product Approvals

The Product Approvals element deals with those products that require approval of their design and construction by some regulatory or other agency independent of the manufacturer. Underwriters' Laboratories for electrical appliances, the Federal Communications Commission and its counterparts in foreign countries for two-way radios, and California's Air Resources Board for automotive emissions are a few of the agencies that come to mind. All pertinent agency requirements must be planned for, as frequently a product needs approval from more than one.

The product design function must provide in its development plan and schedule for the necessary supporting documentation, test reports, samples for agency test, and required validation test participation. Product Design, Marketing, or some administrative function must provide certificates and make logistical arrangements in time for the needed approval before product release for manufacture and sale. The Quality organization is often involved in the testing and certification aspect of the Product Approvals effort, as well as assuring proper marking or placarding of the approved product; but regardless of which organizations do what, the resulting "seal of approval" becomes the basis for customer acceptance of the product as a saleable item.

An important consideration is that a product, to be accepted in the marketplace, must reflect at least the same level of performance, durability, and safety as the test samples on which agency approval was granted. Even if the agency does not police this continuing level of performance, the organization has both the legal and ethical responsibility to assure itself that its ongoing production output is of that high a degree of conformance to specification.

Conducting its new product programs in a manner that will earn the required Product Approvals and will maintain the product's level of performance, a manufacturer will achieve marketability of products in a technical sense, at least, and will minimize certain types of product liability exposure.

2.16 Quality Inputs to Make-or-Buy Decision

Virtually all manufacturing and many merchandising organizations are frequently faced with deciding whether to produce a specific item themselves or to purchase it from an outside source. In some instances, the decision may be phased—initially produced, then procured (or the reverse)—depending on volume changes with time or early availability of certain facilities. Because of the complexity of these situations, most such organizations establish a set of criteria that must be tested in each case to provide the basis for decision. In the Quality System these criteria are set by Quality Inputs to Make-or-Buy Decision.

The criteria involved are either generally applicable or unique to the enterprise; often both kinds are needed. Generally applicable criteria include relative cost of the alternatives, available capacity and comparative delivery, return on investment, inventory and storage costs, and relative quality capability. Unique criteria might include divulgence of proprietary information, the need to improve in-house design capability, and product life. The policies of the enterprise often enter and may be the deciding factor: A policy may demand full "vertical integration," or none; another might identify businesses in which the company "should" or "should not" be, regardless of other circumstances.

In any case, the enterprise would have its own list of criteria and action or decision levels it would exercise in reaching a decision. This element, assuming that the other subjects are covered outside the Quality System, is concerned only with the relative quality capability of vendors and production.

To support this criterion, information produced by 3.2 Supplier Surveys and Capability Determination, 3.10 Supplier Rating Plans, 5.3 Received and Produced Quality Data Reporting, and 6.5 Machine and Process Capability, and perhaps by other elements must be examined to explore comparisons of quality results for similar items. Often, similar items may have been produced by or for competitors or for unrelated applications. Proper validity judgment of the results is then a major ingredient in arriving at the correct decision. As carefully as practicable, the relative quality capability of supplier(s) and production should be determined and quantified. Quantification includes measurement or informed estimate of both quality costs and quality levels or rates of nonconformance. Such measures include the effects of potential shipping, storage, and handling damage.

The satisfactory accomplishment of this effort and the proper use of the results in the decision process will minimize potential future quality problems with the item subjected to the make-or-buy decision.

Subsystem 3

Purchased Material Control

This subsystem, Purchased Material Control, is intended to minimize quality, cost, and delivery problems with suppliers and their products, to the advantage of the supplier, the buyer, and the ultimate customer.

The success of most manufacturing organizations is considerably dependent upon the quality of purchased material. Thus many of the activities of this subsystem and that of Subsystem 4, Process Development and Operation Control, are essentially identical, at least to the extent that they are covered in this book. (The details of their operation in different parts of the factory may well be different, however.) To avoid duplication, these activities will be covered in only one subsystem. Several such activities are covered in Subsystem 3; when the greatest part of the activity is performed elsewhere than in incoming, the activity is described in Subsystem 4.

One of the elements of this subsystem that applies to other subsystems is 3.9 Nonconforming Material Disposition. Many companies charge suppliers who ship them nonconforming material with some portion of their costs of dealing with such material. This is usually done for two purposes: to offset the expense and to punish the supplier. The latter purpose is incompatible with the Quality System's fundamental concept of treating a supplier as a remote part of your own operations (see 3.7 Quality Engineering and Training for Suppliers).

The aim of recovering costs also needs to be examined carefully. Most often the supplier applies such chargebacks to factory or sales overheads, and at a later stage the customer finds his purchased material costs increasing, with appropriate profit amounts included. Competitive conditions may defer such cost increases, but the company that buys much of its material "sole source" rarely profits in the long run from the routine chargeback policy. The policy may be of benefit in special cases, such as with large single lots or high unit-cost items or where a supplier is being permanently disqualified.

The elements of this subsystem are arranged in the likely order of their becoming visible to the supplier. As indicated earlier, some of the activities associated with Purchased Material Control are covered by elements assigned to other subsystems, notably Subsystems 4 and 5.

3.1 Quality Information Package for Suppliers

Most manufacturing companies and many service organizations and institutions are heavily dependent on their suppliers for the ultimate satisfaction of their customers. Suppliers directly affect the quality of products. Despite this vital dependence, far too many organizations do not inform their suppliers adequately of exactly what is needed to ensure such ultimate satisfaction. It should be obvious that an uninformed supplier has a higher potential for quality errors than an informed one. Yet we often treat the supplier-purchaser relationship as an adversarial one in many ways, including withholding needed information through an ill-conceived concern for "security" or from simple oversight.

Thorough communication of all pertinent information to the supplier is the key to good quality in purchased material. When we realize that the optimum control of received quality occurs not from intensive and expensive incoming inspection but from the supplier shipping only "good" material, it becomes apparent that it is to our advantage to provide the supplier *all* the information needed to produce such results. This element is oriented to ensuring that the supplier receives this information at the most useful times—when quoting and when the order is placed—and with appropriate updating as changes occur.

The information should include drawings and specifications, in addition to written, visual, and physical standards. Required inspections and tests to be performed before shipment, as well as those to be performed after receipt, should be detailed, along with statements of the methods, equipment, and work standards to be applied to all related inspections and tests and quality data, reports, and records to be provided with the shipment. Item identification requirements, item preservation, packing materials to be used, packaging testing and marking, control methods and limits to be applied by the supplier or purchaser during manufacture and appraisal of the item, and any special information the supplier might find useful in meeting quality requirements should be given.

Small suppliers often do not take the time to familiarize themselves with all of this type of information given them by purchasers. They frequently do not even read the quality "boilerplate" typically printed on the back of the purchase order or generated selectively for each supplier. Therefore, this element is made more effective when those people carrying out the work of 3.2 Supplier Surveys and Capability Determination and of 3.7 Quality Engineering and Training for Suppliers include a review of the information details with key supplier personnel as part of their normal activities.

When these and the other supplier-related programs of the Quality System are carried out effectively, not only does the level of received quality improve, but many factory and field problems, unrecognized as having a supplier origin, disappear. To make all this happen, we must inform the supplier just what we need—through the Quality Information Package for Suppliers.

3.2 Supplier Surveys and Capability Determination

Many organizations, especially since the early 1950s, have invested extensive resources in attempting to obtain material of satisfactory quality from their suppliers. As the typical manufacturer buys both direct and indirect materials from many sources, with an even larger number of sources as potential suppliers, the ability of all of them to produce quality products on schedule is both very important and an enormous task to determine in advance of production.

One aspect of the supplier's ability to meet the purchaser's requirements is the level of technological sophistication of his production and appraisal equipment. This consideration affects the supplier's costs, quality, and delivery capabilities and must be carefully considered during any evaluation of material sources.

When the material needed for production arrives late or is rejectable for quality reasons, it is too late to find out that the supplier is incapable of doing the job satisfactorily. It is thus to the purchaser's advantage to satisfy himself, in advance, of the supplier's ability and willingness to perform as required. There are several things to consider in making this judgment:

If the supplier has supplied us with similar items, what can we learn from his supplier rating?

Has the supplier supplied similar items to another part of our company or to other companies whose evaluations we trust and can obtain and, if so, what are the results?

Is there business, industry, or government rating available on this supplier, and what are the results?

Can we visit the supplier, observe his operations, talk with his people, study his equipment, evaluate his Quality System, and otherwise satisfy ourselves directly or through agents that he can meet our requirements?

Answers to the first three questions are often viewed as providing supplier capability measures, but to be complete they often must be supplemented with at least an abbreviated supplier survey.

Although each of the listed approaches, singly or in combination with others, can provide the information the purchaser needs, demonstrated performance is both the most economical and the most accurate indicator when it is available. Unfortunately, that is often not the case, and the company must rely on supplier surveys to provide a measure of the supplier's potential to satisfy the requirements. This element concerns itself with those situations.

None of the listed approaches assures that the supplier *will* meet our needs, and this is especially true of supplier surveys. They merely identify those who *cannot* or, as indicated by somebody's previous experience, *will not* or who *may* be able to do so. Many suppliers with an acceptable survey result have failed to perform, and this has led many people to lose confidence in supplier surveys as a valid decision guide. Often, however, their failure to perform has resulted from situations not evaluated in the survey or from other inadequacies in the purchaser's side of the activity. When the Quality System is fully operational, minimizing the purchaser's failings, supplier surveys are a very useful device for identifying potentially good suppliers.

To perform a supplier survey properly requires considerable preparation and, in many ways, is similar to the conduct of an audit (see 1.14 Audit of Procedures, Processes, and Product). The surveyors should become familiar with the drawings, specifications, quality plans, quantities, delivery requirements, previous quality history, and other considerations with respect to the anticipated purchase. A plan, including a checklist, should identify specific areas of the supplier's operations to be examined. Since several areas must usually be evaluated, consideration should be given to having a multifunctional team of experts conduct the survey. Planning meetings and survey rehearsals are necessary parts of the preparation when teams are involved. They are also extremely useful in providing guidance for a single surveyor, particularly if he is an agent rather than an employee of the purchaser.

Insofar as practicable, the items to be evaluated should correspond to and support those included in the formal Supplier Rating Plan (3.10) to which the successful supplier later will be subjected. Many companies develop a supplier survey or capability index that is compatible with and can be converted to their supplier rating indices. Such an index helps Purchasing decide between two or more potential suppliers, one of whom may be a current supplier, on at least a partly objective basis.

The professional conduct of a supplier survey or internal capability analysis produces a report supporting the appropriate index and an identification to the supplier and to Purchasing of what must be done to make the supplier an acceptable source. Of course, both full qualification and disqualification are possible results of the study, as is an incomplete qualification. The report must also guide the planning for incoming inspection and test if the supplier is rated as either fully or partly qualified.

When this program is operated in conjunction with the related Quality System activities of 6.1 Classification of Characteristics, 1.12 Corrective Action Program, 3.7 Quality Engineering and Training for Suppliers, 3.1 Quality Information Package for Suppliers, 3.5 Supplier Control at Source, and those already mentioned in this discussion, the company will realize significant improvements in received quality and associated costs, since Purchasing's supplier selections will have the greatest probability of being correct.

3.3 Supplier Measurement Compatibility

Many problems arise from lack of agreement between factory and field equipment, methods, and standards of inspection and test (see 9.4 Field and Factory Standards Coordination). Organizations encounter similar difficulties when their incoming inspections and tests differ from those of their suppliers, and strenuous efforts are often made to prevent such inconsistencies. Written and physical standards, test fixtures, special gauges, inspection and test instructions, even calibration service (see 7.6 Equipment Calibration and Maintenance) are provided by companies to their suppliers to minimize the potential problems that arise from appraisal incompatibilities.

Suppliers producing complicated components have been known to manufacture parts to pass their customers' gauges rather than to meet their own higher standards. And sometimes this is actually what the customer wants. The Quality System approach is to make sure that such gauges accurately reflect the drawing requirements, that the decisions of the supplier's gauging will be repeated by the customer's, and that *then* the supplier is to produce items that pass both sets of gauging.

Where duplication of equipment or physical standards with the supplier is impracticable, special tests should be run to demonstrate uniformity of decision between the different sets of equipment. Nonuniformity should be corrected. To prevent gradual divergences from causing problems, such correlation studies should be repeated regularly (semiannually is typical), and they should include some items tested previously.

Properly conducted, the activities that ensure the achievement of Supplier Measurement Compatibility with the incoming inspection and test of the buyer will result in reduced time and monetary losses associated with purchased material. These advantages greatly overbalance the investment in planning, tooling, equipment, and services normally required to implement the program.

3.4 Qualification of Parts and Suppliers

Manufacturers, merchandisers, service enterprises, and many government agencies have, to varying degrees, the same problem: They are often heavily dependent on their suppliers for their own capability to perform satisfactorily in the marketplace. In addition to prices charged by the suppliers, the dependence is reflected in the manufacturers' ability to meet promised delivery dates and volumes consistently, their adherence to quality and safety requirements, and the effects of any delivery or quality lapses on the operations of the buyer or of the ultimate customer. For most organizations, such dependence at any given time may be reducible, but it can be eliminated only rarely.

If the relationship between supplier and buyer is to be a continuing fact of life, the buyer, in his thinking and practices to optimize his costs and benefits, is well advised to recognize the dependency. As expressed in 3.7 Quality Engineering and Training for Suppliers, the most effective approach is for the buyer to treat the supplier as an extension of his own operations.

There are numerous aspects to such treatment, as noted in other elements in this subsystem, but among those most directly related to the product or service provided by the buyer is this one, Qualification of Parts and Suppliers. In qualifying suppliers, all available information—from audits, surveys, ratings, trade reports, business publications, and so on—should be used to indicate the supplier's potential to perform satisfactorily. If the evaluation yields a positive result and if sufficient internal experience is included to support it, the decision would be that the supplier is qualified. Lacking such a degree of internal experience in a similar situation, the decision would be a probationary qualification—to be resolved after sufficient material had been considered to support full qualification or disqualification.

Parts qualification starts with the selection or design of the part to serve a clearly defined purpose in the final product, process, or service. The part must have the theoretical capability of satisfying all form, fit, function, and reliability requirements reasonably expected to be placed upon it. Experimental and production parts would then be tested to verify that they meet all the requirements. Such testing, which might require the development of special test programs and fixtures, might include performance, environmental, life, accelerated life, and system application evaluations. Once all the testing is satisfactorily completed, the part can be placed on the qualified parts list. Superb performance under one set of conditions is no guarantee that a part will do as well in others; therefore, special note should be made of the test conditions against which the part was qualified; other applications might require additional tests.

The qualification of a part frequently involves simultaneous qualification of its supplier. As a result, many organizations maintain a qualified supplier/part list to provide a strong basis for valid purchasing decisions and to encourage engineering parts selection practices that emphasize parts standardization and product reliability.

The opposite side of the qualification coin is disqualification. When a qualified supplier fails to perform satisfactorily, the odds are that the buyer has contributed to that failure and may be wholly responsible for it. A thorough examination and indicated correction of the buyer's activities—followed, if necessary, by comparable attention to the supplier—will usually resolve the problem. If the supplier is at fault and is unwilling or unable with all possible assistance to eliminate the cause of the problem, then a new supplier should be qualified and the offending one disqualified, removed from the qualified supplier list—at least, for that part—and restricted in future business with the buyer. In keeping with the "supplier-buyer team" concept, disqualification should be viewed as a last resort.

Qualification of Parts and Suppliers is strongly prevention oriented. Successfully conducted, it contributes significantly to the smooth flow of sufficient parts of proper quality to the buyer. This minimizes delays, work interruptions, and quality problems in the buyer's operations, while reducing associated field or customer difficulties.

3.5 Supplier Control at Source

Since the Quality System is based on preventing quality problems from ever occurring or recurring, it should be apparent that virtually all organizations could benefit significantly from receiving *only* satisfactory items from their suppliers. This would pay off in reducing the amount of incoming inspection, eliminating production delays or rescheduling costs resulting from rejection of purchased material, minimizing factory and field problems caused by supplier items, and other less easily definable benefits to the purchaser.

Other Quality System elements critical to achieving this desirable objective include 3.1 Quality Information Package for Suppliers, 3.2 Supplier Surveys and Capability Determination, 3.3 Supplier Measurement Compatibility, 3.4 Qualification of Parts and Suppliers, and 3.7 Quality Engineering and Training for Suppliers. Many other parts of the System also contribute to this end. The role of Supplier Control at Source must be viewed as covering just those ongoing activities at the suppliers' locations performed by the purchaser to ensure the receipt of only good material.

Many customers, particularly government agencies, maintain at a major supplier's facility full-time personnel who continually audit the process control and product inspection and test activities of the supplier, and sometimes physically inspect or test the product before releasing it for shipment. Industrial customers who perform source inspection do essentially the same thing—frequently on less of an "adversary" basis than is often the case when the government is the customer. Since the intent of the Quality System approach is to establish a mutually beneficial relationship between the buyer and seller, it is important that these programs be developed to help the supplier do the quality job.

Techniques included in this element are source inspection and supplier visitation or surveillance. Source inspection usually is interpreted to mean actual inspection and test of the material by the customer's representative before shipment. Sometimes that requirement may be satisfied by witnessing the appraisal work of the supplier's personnel. The work may be done by full-time personnel stationed at the supplier's location or by visiting agents who perform the inspection or test on a schedule or "on call" before shipment. It is typically viewed as an expensive approach and, therefore, to be applied only in critical situations.

Supplier surveillance may include some product appraisal, but this is often not a lot-release device—just part of the overall evaluation of the effectiveness of the supplier's Quality System in operation. A key feature of the successful surveillance program, as in any audit approach, is the report to supplier management just before the observation tour is completed. Correction of any observed deficiencies should be undertaken as part of the visit, and the discussion with management is intended to ensure that such corrective action takes place.

Personnel performing either source inspection or supplier surveillance should be carefully selected and thoroughly trained not only in the technology required for their activities but also in their responsibilities as agents and for dealing constructively with supplier personnel. Deliberate training of this nature is necessary for all such representatives, but it is of paramount importance when contract services are used for this purpose.

Many companies, particularly smaller ones, when dealing with a geographically remote supplier, have selected a contract service as the most economical way of providing the necessary assurance of supplier quality. By this means, the expenses of travel and of supporting full-time personnel when only part-time effort is needed are minimized. It is also possible to enjoy the benefit of more highly skilled assistance than one could afford to obtain full-time for such effort.

Some supplier quality problems arise because of the customer's purchasing habits. Frequent small orders often contribute to these problems by requiring corresponding setups, whereas a single long-term contract with controlled shipments can often minimize such difficulties. It may also be desirable for the customer to provide financial assistance to a small supplier to enable him successfully to prevent or correct a quality problem.

The data generated by source inspection and supplier surveillance should be compatible with and appropriately factored into the Supplier Rating Plan (3.10) ratings. Thus, the effectiveness of the supplier's use of these programs will contribute to his continuation as a supplier to the company.

To maintain objectivity on the part of personnel working with suppliers, it is advisable to transfer them after some time on the job. The time limits are quite variable but range from a rather unusual six months to about three years, the determinant being the likelihood of the agent's becoming more a part of the supplier's organization than of the buyer's. For this reason, it is necessary to evaluate the effectiveness of the program on a regular basis. Indicated adjustments to the program should be made from this evaluation.

Sometimes these programs, in conjunction with the others mentioned earlier, fail to achieve the results expected—often because our purchases are but a small part of the supplier's business, and we do not obtain the constructive attention of supplier management. Some people have dealt with this situation by marshalling the joint purchasing impact of sister divisions. Such a united-front approach can be effective. When the inattentive supplier is a sole source, creating a competitor is also often successful in getting his attention or in otherwise solving the problem.

These latter tactics are of an adversarial nature and usually are less effective and more expensive and time-consuming than the cooperative approach. Therefore, they should be reserved for those cases where the preferred programs fail.

3.6 Supplier Certification and Objective Evidence of Quality

Large parts of many Quality organizations exist solely to deal with suppliers, reflecting the fact that the related production groups are heavily dependent on those suppliers for their ability to produce at scheduled volume and quality. The question of supplier-customer interdependence has been dealt with in other elements in this subsystem; this element covers the situation that should exist when both sides have a strong commitment to produce only nonconformity-free material.

When that commitment prevails and when the supplier has demonstrated a continuing capability to satisfy the buyer's requirements fully, the opportunity exists for a mutually beneficial approach to the assurance of purchased material quality, called Supplier Certification and Objective Evidence of Quality. In this process, the supplier assures himself of the outgoing product quality through effective process controls and any necessary final inspection and test, and then formally certifies that quality in writing to the customer. Sometimes the supplier provides documentation of the appraisal efforts results—test reports, statistical analyses of inspection and test data, and the like.

The customer then has several options:

To place the material into stock immediately upon receipt

To check just for shipment damage before placing the material into stock

To analyze the objective evidence for undesirable trends or potential problems

To inspect or test the material as would be done without certification (in any case, this should be done in an infrequent random pattern to validate the supplier-provided appraisal data)

The advantages of such a program to the customer are obvious: reduced incoming inspection costs and delays, fewer production problems from rejected lots or defective parts entering stock, and improved supplier relations. The supplier achieves greater assurance of continued sales (often of a larger share of the available market) and a reduced likelihood of shipment rejection by the customer.

3.7 Quality Engineering and Training for Suppliers

During the course of 3.2 Supplier Surveys and Capability Determination, as the result of the application of 3.10 Supplier Rating Plans, and in response to supplier-related problems perceived anywhere within the organization or in the field, it sometimes becomes apparent that individual suppliers or groups of them cannot perform or are not performing satisfactorily. When this occurs, many purchasing managers react by attempting to replace the offending supplier as rapidly as possible. Often this action trades a marginally acceptable supplier for a new but totally incapable one. The economics of this tactic are normally unfavorable and occasionally disastrous.

The approach that has proven most effective is to view the supplier as a somewhat remote portion of our own operation. If we apply to our suppliers the same philosophy we use in dealing with our employees—that everyone would rather do a good job than a bad one, all else being equal—we will find the results comparable to the internal ones. This means that we must commit ourselves to providing the supplier with all the information, standards, and in some circumstances even the tooling needed to meet our requirements, just as we do for our own operations. It also means that we should support the supplier with necessary specific and Quality System training; in effect, we should help set up a Quality System of sufficient sophistication to enable the supplier to improve quality and related efficiencies and thereby satisfy our needs. This, of course, also benefits and gives the program considerable attractiveness to the supplier.

Many people fail to realize the investment a company has in its suppliers. Besides the obvious dependence on the supplier for vital materials to keep its own production going, the typical purchaser has invested significant Purchasing, Development Engineering, and Quality departmental money in the supplier and has often advanced funds for or has directly provided raw material, tooling, equipment, or even facilities. But more than all these is frequently the transfer of information from the purchaser to the supplier. Not only has the purchaser invested time and materials in providing the supplier with drawings, specifications, inspection plans, standards, purchasing guides and purchase terms, and all the other specific and general contributors to "how to do business with us," but the supplier has usually invested much time and effort to assimilate such information into his working situation. The purchaser unwittingly pays for all this supplier effort as part of the price of the purchased material; and the investment, particularly as trade continues, continues to grow, albeit at a slower pace than initially. To throw all this away and face another large investment with a new supplier fails to recognize that the old supplier has a "leg up." If we can turn him around, it will normally cost less than going elsewhere and starting all over. Sometimes it will be necessary to develop a new supplier rather than throw good money after bad, but these times will be rare. We should be prepared to do this whenever required, but not as the first action choice.

Another important advantage to the purchaser in this entire approach is that suppliers will be far more loyal when treated constructively as "members of the team." In times of material or capacity shortages, "good" customers often receive preferential treatment. And good does not always mean large; what it often means is cooperative, supportive, and positive. By providing Quality Engineering and Training for Suppliers, a company can ensure its reputation within the trade and can expect to be viewed and treated favorably.

In carrying out this program, we must be prepared to provide the supplier with support comparable to what we would do for an internal department in the same straits. We would train people—operators, inspectors, engineers, and managers alike; we would make capability studies; we would improve work layouts, practices, and tooling; we would initiate permanent, effective, corrective action; we would change specifications; and we would do anything else needed to ensure the success of our investment in this supplier. Timeliness is often important, so we would even empower our quality engineers to make certain decisions and commitments while working with the supplier.

A problem sometimes arises between a small customer and a large supplier in applying this element. The supplier may well have more extensive and even more capable personnel resources than the purchaser, and the disparity might be so great that only embarrassment would result if the normal approach to this element were undertaken. Such a situation does not preclude useful contacts, however. The supplier may not be applying sufficient energy to serve our needs, and a visit with pertinent discussion could well cause such effort to be expended. There have been only two or three instances in my experience when a supplier has refused to act upon a valid request constructively presented.

Whether or not the situation requires an actual visit to the supplier's facility, the quality engineer or other person(s) involved should prepare for the contact by becoming fully informed about the specific problem requiring attention. This can be as extensive as preparing checklists and action plans, or it can be as limited as studying the involved part and its quality requirements and history. In any case, the quality engineer must provide the supplier with accurate, thorough, effective support.

After his work with the supplier is complete, at least for this contact, the quality engineer must produce a timely report to the supplier, Purchasing, and incoming inspection management. The report should describe the problem, the action taken, and actual or anticipated results. If formal training was provided to the supplier, the subjects covered should be listed as well as the specifics of any key contractual interpretations or legal considerations, such as product liability matters. A record is necessary also when the purchaser conducts similar training programs with groups of suppliers. It is important, if challenged, to be able to substantiate that such critical training has been provided.

The benefits to the purchaser from providing this kind of service to suppliers are significant: Costs are reduced, quality is improved, time is saved, and delivery schedules are met.

3.8 Appraisal Status Marking

Parts, assemblies, and final product often go through a number of inspection and test operations in the process of receipt and further manufacture. Considering the volume of material handled and the diversion, for a variety of reasons, of some of that material away from the rest, it often becomes vital to provide for carefully controlled marking on the items or on their accompanying paperwork. The marking should indicate just which appraisal operations have been completed and what the results were.

This is done by stamping, punching, etching, or otherwise applying a permanent mark to the item when a full-blown Material Tracking (1.9) program is applicable and practicable. If such measures are not required or are not feasible, accompanying route cards, tags, or logs may be appropriately marked and retained for reference as described in 1.8 Product Liability Prevention. Often each inspector is issued controlled stamps bearing a unique identifying code, within symbols showing the type of inspection performed and the disposition of the item. The item or related document then displays at any stage a series of stamps providing a sufficient summary of its quality history up to that point.

Such a record tends to prevent omitting required rework, repair, reinspection, or scheduled appraisal operations. It also serves to prevent later manufacturing operations from being performed prior to needed rework or on material scrapped in process. There is also a psychological benefit, for the inspector or tester feels a positive identification with and responsibility for the use of the unique stamp.

3.9 Nonconforming Material Disposition

The Nonconforming Material Disposition element provides the formal procedure for identifying, storing, and processing nonconforming material expeditiously and for preventing known nonconformances from reaching the customer. When recognized, nonconforming material, whether received from a supplier, from a previous operation, or from another department, creates the problem of dealing with it. This involves the determination of just what constitutes nonconformance, who has the pertinent responsibilities, what is to be done with the nonconforming items, what is to be done with the conforming items, how both groups are to be identified and separated, what records are to be initiated, what importance is to be attached to various types and degrees of nonconformity (see 6.2 Classification of Nonconformities), what corrective action is called for, and who is to be informed. If all these details must be freshly decided in each case, the aggregate impact on most organizations would be prohibitive. Therefore, a formal approach is necessary.

What constitutes nonconformance is determined by drawings, specifications, standards, practices, negotiation, contracts, and other sources. The application of the resulting criteria to the decision process and their useful interpretation require careful planning and dissemination to all involved. The use of demerit schemes for rating the seriousness of various levels of nonconformance is covered in 3.10 Supplier Rating Plans.

The answer to the responsibility question is to examine each step in the disposition process and determine which function or functions have the basic charter associated with the activity and can most readily perform it while satisfying the system concepts in the process. For example, the organization that detects a discrepancy—whether it be incoming or final inspection or test, line operations, or field personnel—has the responsibility to report it expeditiously to the involved agencies, including suppliers, and to identify and segregate the material in accordance with stated practice. The design agency normally has the responsibility to decide the disposition if the discrepancy affects form, fit, or function, whereas Manufacturing decides the disposition in cases affecting further processing. But in all these decisions there is a necessary involvement of production control and the Quality function and, conceivably, product safety personnel. And if the discrepancy is found in the field, service personnel are also involved.

For this reason, the use of a material review board or committee has become rather widespread. A typical arrangement would be to have a screening group empowered to order certain dispositions, after verifying that the condition is really as reported. The disposition could include return to supplier, scrap, salvage, reclaim, process further and eliminate the condition, or rework. If the screening group could not agree or felt that other dispositions were applicable in a given instance, that case would be referred to the material review board. The board is composed of senior management personnel and, sometimes, customer representatives. The board could make additional dispositions such as repair, change the design to conform to the product, "use as is," downgrade, or any other appropriate decision. In some specific cases the board may delegate to the screening group certain of these decisions, with necessary controls. It is important that the notification, screening, board action, and disposition processes be actively pursued and expedited. Delays influence the dispositions improperly, damage occurs, costs mount, and the potential for useful corrective action is reduced as time passes.

Dispositions are normally applicable only to the items found discrepant or to the entire lot from which they came. Special arrangements and approvals usually are required to apply the dispositions, particularly those involving acceptance of discrepant items or additional work, to other lots. The extent of the disposition and the basis for accounting for extra work should always be made clear, as should any required reinspection or retest of the product.

Once the disposition is made and documented, the decision should be distributed to those involved and, to avoid the undesirable effects of delay mentioned above, implemented as rapidly as practicable. Rapid implementation is also desirable to achieve the earliest feasible assessment of the correctness of the decision, thus permitting countermeasures that may still allow meeting schedule requirements. Close follow-up is necessary to confirm that all decisions are carried out on schedule.

It is imperative that the entire Nonconforming Material Disposition process be reviewed regularly to ensure that it is working effectively. One key consideration in the review is the prevention of known nonconformities being passed to the customer. This is a major contributor to 1.8 Product Liability Prevention. It is also important to examine the data generation, processing, and reporting associated with nonconforming material to verify that proper forms are used; that nonconformity patterns can be studied by supplier, product, operator, department, or by any other useful category; and that the program is economical but complete.

The aim of the program to control nonconforming material is rapid action to minimize the production of a large volume of similar nonconforming items. This saves money and time and eases pressure to accept serious nonconformities solely because they exist in large volume. Consequently, this minimizes the likelihood of including such items in customer-bound product.

3.10 Supplier Rating Plans

Many companies have adopted Supplier Rating Plans to enable Purchasing Department personnel to make the best overall decisions on awarding contracts and to help identify those suppliers needing priority assistance on quality problems. In these plans, attempts are made to incorporate, properly weighted, all of the considerations that actually should influence these decisions. In addition, the ratings can be used as a positive recognition of supplier quality accomplishment—through a "Supplier of the Month" award, for example.

Simple or complex, the plans usually include at least some measure of the known Cost, Quality, and Delivery experiences with the supplier. Other plans include aspects of supplier cooperation in key programs, often called a Service factor. But whatever the plan type, points are awarded or deducted, indices are calculated, and relative rankings of the suppliers are made.

In some organizations, Purchasing and support decisions are fully controlled by the supplier ratings; in most, these decisions are given some leeway within the guidance of the ratings. The Quality System requires that the developed plan and ratings properly reflect the real criteria for decision and that any decision outside the ratings be documented and approved in advance of implementation at a proper System management level.

A useful approach to developing a rating plan provides separate indices for each key factor, with an overall index that combines them. Thus one index would be developed for the received quality of the suppliers' shipments; another would cover cost, usually involving a comparison with competitive quotes; a third would show the pattern of receipts versus promised delivery dates; a fourth would measure the service considerations appropriate to the supplier situation; and then there would be the overall rating index, combining the others according to some formula.

The least applicable index many times is the cost factor. Many purchases are made from "sole sources," and no competitive data are available; or much material is purchased from other divisions of the same company, and price competition is not a working basis for a purchasing decision. In these situations, the cost factor is omitted from supplier rating, although the Purchasing department personnel obviously consider price whenever they can.

Since the conditions being measured are all to be compared with some goal, the choices of number patterns are numerous. Some people use percentages, others measure deviations from a goal, others construct demerit rating schemes with a one-sided limit on scores, and still others combine two or more such methods into their rating practices. Many companies have found useful an approach that has each index fluctuate around 1.000. The advantage of this tactic is that, if the scales are properly set, the index can serve as a multiplier of bid price, thus reflecting the real costs of doing business with competitive bidders. Such an application, of course, requires the omission of the cost factor from the total index; otherwise Cost would be involved twice in the competitive comparisons.

An illustration of a rating plan using the 1.000 base follows. It is important to recognize that many other useful approaches to supplier rating exist. This is just an example of a rather complete one. Alternatives are discussed later in this element. The indices to be generated in this plan are for Quality, Delivery, and Service—with a composite total of these three, properly weighted. The weighting factors, usually involving Purchasing, Engineering, Manufacturing, and Quality, must be internally negotiated, but often develop patterns such as Quality = .55; Delivery = .25; Service = .20. When Cost is another index, the pattern is often close to: Quality = .40; Cost = .35; Delivery = .15; Service = .10.

The Quality factor should contain a measure of lot-by-lot nonconformity levels, a severity component based on 6.1 Classification of Characteristics and 6.2 Classification of Nonconformities, and a further component reflecting the final disposition of the lot. Included in the nonconformity determination should be questions of completeness of the order, proper packaging, and support documentation.

The Delivery factor recognizes, from a handling, storage, and carrying cost standpoint, the undesirable nature of premature delivery, as well as late-delivery disruptive effects such as expediting costs, rescheduling of production, and customer irritation. On-time delivery is the only wholly satisfactory condition, although one would normally prefer to receive the material early rather than late. Of course, if we are involved in a Just-in-Time (JIT) effort with our suppliers, early delivery is no more acceptable than late. In that case, the demerit scale in Table 2 would have to be changed. Since a shipment may contain a mixture of scheduled items (such as half of it ten days late and the remainder current), the demerits awarded should reflect that mixture (e.g., in the above example, the late half would earn a .100 rating and the current half a 0, for an average of .50).

The Service factor is the most variable as to content. Selected for inclusion in the index should be only those supplier actions and responses that are significant in the context of each user's Quality System and what that supplier is expected to provide. Examples include, but are not limited to, the following:

Cooperating in pre-award surveys
Acknowledging requested delivery dates
Exhibiting concern for personnel safety in product and packaging characteristics
Providing preproduction samples and quality plans
Providing process classification of characteristics information
Providing advance notification of process changes affecting quality
Using only the most capable methods and equipment to produce required product
Providing notice well before shipment of any known lot-quality problems
Taking and reporting effectiveness of corrective action on quality problems
Participating in ongoing quality improvement efforts
Accepting and using outside quality engineering and other technical help to improve the effectiveness of the Quality System, when offered
Reacting effectively to requested schedule changes
Participating, when requested, in a JIT program

A pattern of merit and demerit ratings for each of these factors is shown in Tables 1, 2, and 3.

The Service factor contains components that may vary by lot or on some other basis, such as by order, by problem, or by inquiry. Therefore, the Supplier Service Index may not change as rapidly as the others. Evaluations for each of the selected components are normally developed jointly by Purchasing and Supplier Quality Engineering. The Supplier Service Index is the only factor in the rating that allows a value less than 1.000. This permits the supplier who is cooperative and responsive to our needs to enjoy an earned preference within the total group of suppliers.

Supplier ratings are ordinarily published monthly or quarterly, with updates on demand. Therefore, the experience of more than one lot of a given part and more than one part from a single supplier are often involved. To minimize confusion in the necessary comparison of ratings among competing suppliers, a uniform approach should be taken. One method fre-

TABLE 1 Quality Factor Demerits

Observed Lot % Noncon-forming	*(L) Demerits*	*Noncon-forming Charac-teristic Classi-fication*	*(C) Demerits*	*Noncon-formity Classi-fication*	*(D) Demerits*	*Lot Dispo-sition*	*(S) Demerits*
0– 0.1	0	Incidental (I)	1	Minor (N)	.001	Accept	0
0.1– 0.5	.01	Minor (N)	3	Major (J)	.002	Use as is	.05
0.5– 1.0	.02	Major (M)	6	Serious (S)	.005	Return to vendor	.10
1.0– 2.0	.05	Critical (C)	10	Very serious (V)	.010	Sort	.15
2.0– 5.0	.10					Rework	.15
5.0–10.0	.20					Repair	.15
>10.0	.50					Scrap	.20

$$\text{Supplier Quality Index} = 1.000 + \sum_{M=1}^{25} \left[\frac{L + \frac{5N}{nb} \sum_{d=0}^{d} (C \times D) + S}{M} \right]$$

Where L, D, and S are the appropriate demerits from Table 1; C is the sum of the corresponding Importance Values, which cannot exceed 10 (or Critical); d = number of nonconforming characteristics in the sample; n = sample size; N = lot size; b = the total number of characteristics inspected in the sample; and M = the number of lots in the Index. The multiplier of 5 in the sample Quality factor is used to weight that factor appropriately with respect to those for lot quality and disposition. A different multiplier could be used, depending on the desired contribution of sample quality to the overall quality demerit total.

quently employed is to use the data from only the most recent 25 lots of a part and to report ratings by part. This permits the most direct comparison among suppliers. When it is useful to look at a supplier's overall performance, however, the 25-lot indices for all or selected parts may be combined, often weighted by volume or relative cost.

TABLE 2 Delivery Factor Demerits

Actual Date vs. Acknowledged Date	*(A) Demerits*
Early >20 days	.050
Early 6 –20 days	.010
Early 0 – 5 days	0
Late 1 – 5 days	.050
Late > 5 days	.100

(See delivery factor discussion on page 122 for mixed-lot treatment)

Lot Delivery Index = 1.000 + A

TABLE 3 Service Factor Ratings

	Service Provided		
Service Characteristics	*Excellent*	*Satisfactory*	*Unsatisfactory*
a. Cooperating in pre-award surveys	+ .050	+ .010	− .050
b. Acknowledging requested delivery dates	+ .020	+ .010	− .010
c. Exhibiting concern for personnel safety in product and packaging characteristics	+ .050	+ .010	− .050
d. Providing preproduction samples and quality plans	+ .020	+ .010	− .020
e. Providing process classification of characteristics information	+ .050	+ .020	− .010
f. Providing advance notification of process changes affecting quality	+ .050	+ .020	− .050
g. Using only the most capable methods and equipment to produce required product	+ .050	+ .020	− .020
h. Providing notice well before shipment of any known lot-quality problems	+ .050	+ .020	− .050
i. Taking and reporting effectiveness of corrective action on quality problems	+ .050	+ .020	− .050
j. Participating in ongoing quality improvement efforts	+ .050	+ 0.20	− .050
k. Accepting and using outside quality engineering and other technical help to improve the effectiveness of the Quality System, when offered	+ .050	+ .020	− .050
l. Reacting effectively to requested schedule changes	+ .020	+ .010	− .010
m. Others (including JIT participation)	(assign values as appropriate)		

Supplier Service Index = 1.000 − [ΣCharacteristic Ratings ≤.200]

On this basis the Supplier Quality Index (SQI) by part would be the sum of the 25 most recent lot quality indices divided by 25. The Supplier Delivery Index (SDI) by part would be similarly calculated. And so would the Supplier Service Index (SSI), if it was prepared on a lot-by-lot basis.

Often, however, that index is developed only monthly or quarterly—to correspond with the normal supplier rating publication schedule—and may cover the supplier's entire performance, not isolated to just one part.

The overall or composite Supplier Rating Index (SRI) would then be calculated as follows, using the weighting factors mentioned earlier:

$$SRI = .55\ SQI + .25\ SDI + .20\ SSI$$

An imaginary example might be helpful. Let's assume that the latest 25 lots of part X from Supplier Y had the history shown in Table 4. Accordingly, the delivery and quality demerits for the 25 lots would be those shown in Table 5. Let us also assume that the most recent inputs in the Supplier Service Index for Supplier Y are as described in Table 6. The Supplier Service Index would then be:

$$SSI = 1.000 - (.050 + .010 + .010 - .010 + .050 - .020 - .050 - .050 + .020 + .020 + .010 + .020) = 1.000 - .060 = .940$$

The composite Supplier Rating Index for this example would then be:

$$SRI = .55 \times 1.303 + .25 \times 1.034 + .20 \times .940 = .717 + .258 + .188 = 1.163$$

The interpretation of such an index would be that the bid price should be multiplied by 1.163, or increased by approximately 16 percent, to give us a clearer picture of our actual costs of doing business with this supplier. Such conclusions, derived from the use of a Supplier Rating Plan for all suppliers, enable Purchasing to buy preferentially from those suppliers who perform in our mutual best interests and to avoid the trap of the "low-price, high-support-cost" supplier.

The example shows the use of lot sampling data obtained from Incoming Inspection records. Companies, such as chemical processors, that do not purchase "pieces" would omit or modify the sample quality factor to reflect its significance in such a situation. Others might wish to incorporate source-inspection results or a revision of the rating as a result of unsatisfactory conditions found after release of the lot to production. All of this can be accomplished by appropriate adjustment of the index formula and by setting up the necessary data collection and processing programs.

Although the supplier rating is intended primarily to be used, as previously stated, to give Purchasing a valid measure of "real" costs as the basis for making procurement decisions, it can also be used with preestablished standards or goals as the basis for qualification or disqualification of suppliers. The addition of suppliers to or the threat of their removal from an approved bidders list can have a salutary effect on their performance. For this reason, it is important to have a formal program for dealing with various

TABLE 4 Example Lot History

Supplier Y | *Part No. X*

Lot No.	*Date Received*	*Date Promised*	*No. Pieces*	*Pieces Inspected*	*Pieces Nonconforming*	*Characteristics Inspected*[1]				*Characteristics Nonconforming*				*Nonconformance Classification*[2]				*Disposition*[3]
						I	N	M	C	I	N	M	C	N	J	S	V	
38	1/7	1/4	850	20	1	40	60	60	100	1	0	1	0	0	2	0	0	Sort
39	1/14	1/14	700	20	0	40	60	60	100	0	0	0	0	0	0	0	0	Acc
40	1/21	1/25	800	20	0	40	60	60	100	0	0	0	0	0	0	0	0	Acc
41	2/10	2/3	700	20	2	40	60	60	100	0	2	1	1	1	1	1	1	Rep
42	2/10	2/14	700	20	1	40	60	60	100	0	0	0	3	0	0	0	3	UAI
43	2/15	2/25	920	20	0	40	60	60	100	0	0	0	0	0	0	0	0	Acc
44	3/4	3/3	1000	32	0	64	96	96	160	0	0	0	0	0	0	0	0	Acc
45	3/11	3/11	1000	32	0	64	96	96	160	0	0	0	0	0	0	0	0	Acc
46	3/14	3/23	800	20	0	40	60	60	100	0	0	0	0	0	0	0	0	Acc
47	3/14	3/31	600	20	1	40	60	60	100	2	3	0	1	1	1	0	4	Sort
48	3/14	4/3	50	10	0	20	30	30	50	0	0	0	0	0	0	0	0	Acc
49	3/14	4/14	400	20	0	40	60	60	100	0	0	0	0	0	0	0	0	Acc
50	4/29	4/29	1150	32	0	64	96	96	160	0	0	0	0	0	0	0	0	Acc
51	5/5	5/5	1000	32	1	64	96	96	160	0	0	0	1	0	1	0	0	UAI
52	5/31	5/13	800	20	3	40	60	60	100	2	5	3	6	0	7	3	6	RTS
53	6/1	5/24	660	20	1	40	60	60	100	0	2	1	2	0	1	3	1	Sort
54	6/7	6/3	820	20	0	40	60	60	100	0	0	0	0	0	0	0	0	Acc
55	6/15	6/14	900	20	0	40	60	60	100	0	0	0	0	0	0	0	0	Acc
56	6/20	6/17	1000	32	8	64	96	96	160	5	6	8	8	1	9	10	7	Scr
57	6/28	6/28	1000	32	1	64	96	96	160	1	1	0	1	0	0	2	1	Sort
58	7/22	7/19	930	20	1	40	60	60	100	0	0	2	0	0	1	0	1	UAI
59	8/2	7/29	1000	32	1	64	96	96	160	0	0	1	1	0	1	0	1	UAI
60	8/4	8/4	1000	32	0	64	96	96	160	0	0	0	0	0	0	0	0	Acc
61	8/12	8/11	1000	32	0	64	96	96	160	0	0	0	0	0	0	0	0	Acc
62	8/22	8/18	1000	32	1	64	96	96	160	0	0	0	2	0	0	1	1	UAI

[1] I = Incidental, N = Minor, M = Major, C = Critical
[2] N = Minor, J = Major, S = Serious, V = Very Serious
[3] Acc = Accept, Rep = Repair, UAI = Use as is, RTS = Return to supplier, Scr = Scrap

index results, including the individual indices of Quality, Delivery, Service, and Cost, if used, and sometimes even their components. Such a program should include the criteria for issuing corrective action notices, for approved bidders list qualification and disqualification, for purchase order

TABLE 5 Example Delivery and Quality Demerits

Lot No.	*Delivery Demerits*	*Lot Quality Demerits (L)*	*Importance Values (C)*	*Nonconformance Class Demerits (D)*	*(C) × (D)*	*Lot Disposition Demerits (S)*	*Total Quality Demerits*
38	.050	.100	7	.004	.028	.150	.278
39	0	0	0	0	0	0	0
40	0	0	0	0	0	0	0
41	.100	.200	10	.018	.180	.150	.530
42	0	.100	10	.030	.300	.050	.450
43	.010	0	0	0	0	0	0
44	.050	0	0	0	0	0	0
45	0	0	0	0	0	0	0
46	.010	0	0	0	0	0	0
47	.010	.100	10	.043	.430	.150	.680
48	.010	0	0	0	0	0	0
49	.050	0	0	0	0	0	0
50	0	0	0	0	0	0	0
51	0	.100	10	.010	.100	.050	.250
52	.100	.500	10	.089	.890	.100	1.490
53	.100	.100	10	.027	.270	.150	.520
54	.050	0	0	0	0	0	0
55	.050	0	0	0	0	0	0
56	.050	.500	10	.139	1.390	.200	2.090
57	0	.100	10	.020	.200	.150	.450
58	.050	.100	10	.012	.120	.050	.270
59	.050	.100	10	.012	.120	.050	.270
60	0	0	0	0	0	0	0
61	.050	0	0	0	0	0	0
62	.050	.100	10	.015	.150	.050	.300
Total	.840						7.578

$$\text{Supplier Delivery Index} = 1.000 + \frac{.840}{25} = 1.000 + .034 = 1.034$$

$$\text{Supplier Quality Index} = 1.000 + \frac{7.578}{25} = 1.000 + .303 = 1.303$$

placement and cancellation, and for the application of support programs such as 3.5 Supplier Control at Source or 3.7 Quality Engineering and Training for Suppliers.

On a regular basis, at least annually, the organization should review the utility of this rating plan. This review should be both objective and subjective, with a thorough evaluation of the net favorable or unfavorable impact of the program on received quality, delivery, and overall cost. Purchasing personnel should also provide supplier reaction inputs and an assessment of weighting factors. If changes to the plan are indicated, they should be made. Of course, revisions to the Service factor should be made

TABLE 6 Example Supplier Service Data

	Service Provided		
Service Characteristics	*Excellent*	*Satisfactory*	*Unsatisfactory*
a. Cooperating in pre-award surveys	X		
b. Acknowledging requested delivery dates		X	
c. Exhibiting concern for personnel safety in product and packaging characteristics		X	
d. Providing preproduction samples and quality plans		N/A	
e. Providing process classification of characteristics information			X
f. Providing advance notification of process changes affecting quality	X		
g. Using only the most capable methods and equipment to produce required product			X
h. Providing notice well before shipment of any known lot-quality problems			X
i. Taking and reporting effectiveness of corrective action on quality problems			X
j. Participating in ongoing quality improvement efforts		X	
k. Accepting and using outside quality engineering and other technical help to improve the effectiveness of the Quality System, when offered		X	
l. Reacting effectively to requested schedule changes		X	
m. Participating in JIT program		X(+ .020)	

when circumstances change; new supplier facilities or personnel can result in a rapid adjustment in some of the components of the Service factor.

The Supplier Rating Plan evaluation is accomplished partly as an ongoing observation, since trends in ratings are normally included in the regular reports. If the rating trends correlate with received quality and other factors, the rating can be considered reasonably valid. If they do not correlate, thorough investigation is required.

In most supplier rating plans, the bulk or all of the data are generated from established suppliers and purchased material; some people even exclude data until at least 25 lots have been received. The plan described here can accommodate any supplier-lot situation mathematically. Most companies, however, take certain precautions concerning a new supplier or a new purchased item. These may involve the required submittal of a qualification sample in advance of production shipments, more extensive evaluation of the first lot received, or inspection visits to the suppliers as they are making the first production runs of material.

When a qualification sample is submitted, it is not usually counted as a lot for supplier rating purposes; the sample and perhaps the resultant quality measurement would be counted as part of Supplier Service Index component d, as listed on page 125, within the eventual rating for that supplier. Such a sample is often carefully measured, generating variables data rather than attribute, for all specified characteristics and the data used for capability estimates and retained for later comparisons. At frequent intervals, often semiannually, the customer company will draw a comparable sample from a randomly selected production lot from the supplier, subject it to the same type of scrutiny as was applied to the initial qualification sample, compare the new data with previous sets, and take any indicated action. Such later sample results would be included in the Supplier Quality Index and possibly also in the Supplier Service Index under component f or h.

This discussion has been based on a fairly complex plan, useful in many situations to discriminate rather precisely among suppliers of similar capabilities. A further complication to the arithmetic and ease of understanding of the program by everyone, albeit one that emphasizes the most recent supplier experience as being a better forecast of likely performance, is exponential smoothing of the lot data or indices. But many organizations, to minimize training requirements and misunderstandings, prefer a much simpler approach. Such an approach may involve the use of just the percentage of nonconforming lots received and the percentage of lots received late, without the use of a demerit rating. The two percentages are then weighted as to relative importance and added together to produce an index usable as above.

Whatever the approach used, the operation of the Supplier Rating Plan in conjunction with 6.1 Classification of Characteristics, 6.2 Classification of Nonconformities, and the previously mentioned supplier support programs will result in continuing improvement in supplier quality and delivery at lowered overall cost to the user of the plan.

One of the outcomes of a Supplier Rating Plan is that it provides a suitable basis for 3.6 Supplier Certification and Objective Evidence of Quality. Another outcome is the establishment of a formal Supplier Recognition Program. Outstanding performance by a supplier can produce plaques, public notices, and other recognition from the pleased customer or other agencies. Such awards can be used by the supplier to support an application for the Malcolm Baldrige National Quality Award, for example. Providing such recognition, then, can have a valuable effect in developing a long-term positive relationship between the supplier and customer companies.

Subsystem 4

Process Development and Operation Control

Process Development and Operation Control logically extends controls over process design and suppliers to produce a minimum degradation of customer satisfaction and profitability resulting from internal manufacturing activities.

As described in the introduction to Subsystem 3, Purchased Material Control, the activities of several elements in that Subsystem and in this one overlap. This is entirely consistent with the approach to be taken in developing the Manufacturing Quality Plan (4.7), which covers the full range of Quality System activities associated with a specific production item in a factory.

One of these overlapping elements is 4.10 Inspection and Test Performance, which is pertinent to either incoming or production inspection and test. In production, however, there is available to the responsible organization an alternative that is rarely considered for incoming. It is feasible to have all or part of the needed in-process and final inspection and test performed by production personnel rather than by independent inspectors or testers.

The rationale for this thinking is that it supports the "unit manager" concept for production management, provides "job enrichment" for certain operators, and minimizes the "policeman" stigma frequently attached to separate inspectors. It also provides a sound basis for applying the philosophy that everyone has "internal" customers as well as (perhaps) unknown "external" ones who must also be satisfied with whatever is produced. This is true, of course, regardless of the functional assignment within the organization.

When all managers adopt the philosophy of the Quality System and pursue their quality responsibilities as diligently as they do those for production, this approach works well and results typically in cost savings and improved morale among operators. Short of achieving such a level of maturity, most organizations find it necessary to modify the approach by having a continuing independent audit of the program. The Quality System already provides for such an audit, so there is no basic disagreement with this type of program, if it is properly controlled.

The elements in this subsystem are arranged as nearly as possible in the order in which they would be applied in setting up a new factory.

4.1 Process Design Review

A product design must be carefully evaluated at several stages during its development, as prescribed in 2.9 Product Design Review. Similarly, a new process accompanying the introduction of a new product or being introduced on its own must also be carefully evaluated. The organizations involved in a process review might be somewhat different from those performing a product review, particularly if the reviews are held independently of each other. But many companies insist that processes be developed simultaneously with the products requiring them and that the reviews be held together; of course, this would not apply for a new process applied to existing products.

Typically, Process Design Review involves the process design group, Production, Safety, the Quality organization, Purchasing and equipment suppliers as appropriate, Product Engineering, and Facilities Engineering. Since the design of a process needed for a new product cannot normally be undertaken until the product development cycle has been through several stages, the process designers are often at a time disadvantage. This frequently results in sub-optimizing both manufacturing capability and costs for a new product for a substantial time period—in worst cases, permanently. Therefore, every effort should be expended to start process design early enough in the product development cycle to make sure that it has the best chance of achieving desired quality and production rates. The timing is controlled by the New Product Introduction Activity (2.1) element, and Process Design Review is aimed at ensuring the full success of the process design effort.

In the example given in 2.9 Product Design Review, the forced compatibility of product and process worked both ways. The process approaches—manual, mechanized, and automated—were mutually evaluated and made wholly responsive to product needs and to those involving ensuring satisfaction of key quality criteria. The processes were also developed to be capable of satisfying scheduled production buildups with increasingly sophisticated capacity and with time available to train operating personnel adequately. This resulted in full production needs being met by fully automated equipment, while lower levels of production were handled manually or with mechanized processes. Without such parallel process development effort, full production demand would have had to be met with less capable and more costly methods. And, as such matters often go, the automated equipment might never have been built, to the cost detriment of the company.

In Process Design Review the same principles hold as for products: Time is allowed for sufficient review, a neutral function chairs the meeting, decisions of the review group are binding on the designer, and design revisions are followed up and properly documented. Obviously, the process should be designed to be basically stable and as insensitive to changes in operating conditions, materials, and personnel as possible; it should produce the item with maximum quality and safety at minimum cost; and it should be easy to operate and control. As the example implies, associated measurement and control equipment should be an integral part of any mechanized and automated processes, thus minimizing appraisal and internal failure costs.

The advantage of reviewing the process design should be obvious from the preceding discussion. Compatibility with the products, provision for coordinated inspection and automatic control, and minimization of manufacturing and field quality problems all result from proper process design and timely design review and correction.

4.2 Process Documentation

For many years, contractors to government agencies that procure high technology items have been required to document their production processes. The documents are provided to the operating personnel and used as the basis for conformance audits conducted by the internal Quality organization and by government personnel. Although many companies initially performed this work of formalizing their processes only because it was a contractual requirement, they have found that it has a salutary business effect as well.

It requires them to do a better job of process planning, thus reducing manufacturing and quality problems implicit in the uncontrolled situation; it provides a training basis for new operators, saving time, reducing turnover, and improving the quality of their early output; and it eliminates quality and other problems associated with shift changes or expansion of the production work force, particularly at remote sites. Many of these organizations, having recognized the beneficial impact of these results, make Process Documentation part of their operating programs even after they have ceased being contractors to the government or when they set up other types of businesses.

One of the key features of process specifications is their requirement for tolerances on process settings. Target values, although useful, provide no standard to be used as a basis for control of a process; how much of a deviation from target requires process correction is a matter for individual judgment. When tolerances are set, the element of doubt is removed. To be fully effective, these limiting values of the process parameters must be established before production begins, setting the stage for process or product redesign to eliminate conflict or to provide direction for resource application control. The values are also subject to verification during early production, with associated corrections as required.

The process specifications must provide direction to the operator on what things to do, on what material, in which order, with what equipment, needed safety precautions, the time required, inspection and test actions, gauges used, and any other pertinent guidance. They must also establish, for line supervision, such details as applicable requirements for ambient conditions, setup controls, intermediate dimensional controls, and waste disposal and other environmental considerations.

Since so many items of information must be transmitted effectively, it is imperative that the program concern itself with the recipients' points of view. Documents should be legible, as brief as possible consistent with understanding, clear, and written for the educational level achieved by the users, while covering all items sufficiently. That these documents produce the intended communication efficiency should be tested. There must also be a provision to update the documents rapidly and accurately with changes in process, product, or work force characteristics. Some manufacturers have successfully used computers to generate or to display this information in production areas.

At the completion of the product and process development cycle, a last look should be taken at the process specifications to assure that they correctly reflect the state of manufacturing knowledge at the point of release for production. This includes any pertinent feedback from the field associated with engineering tests or pilot-run experience. Later revisions will occur as production experience develops, but production must start with the best level of training and control documentation possible. This offers the best chance of achieving volume and quality goals early in production.

For continuing success in production, the documentation-of-process step is sufficiently important that it should be accomplished by any appropriate means. In the absence of available technical personnel, many companies have assigned this responsibility to line supervision and experienced operators. The contribution by the operators can be very significant and has the added benefit of motivating them to use the resulting documented process as a matter of personal pride.

4.3 Special Process Control

Many manufacturing sequences include operations that cannot effectively or safely be performed by all members of the work force. These operations require specialists with varying degrees of necessary formal training to operate the unique equipment and, often, to evaluate the results. The lists of such operations often are different from one company to another even in the same industry but are usually identified as "special processes." Some of the processes that have carried this designation over the years are metal joining (welding, brazing, soldering), heat treatment, painting, plating, and nondestructive testing.

Practitioners of these techniques are often of a category different from other factory personnel; they may be certified by unions, their companies, educational bodies, technical societies, or government agencies, and they enjoy commensurate benefits. This subject is covered further under 8.11 Qualification Standards. In addition, the technology involved in their crafts may be of a type or degree beyond the capabilities of their organization's technical staff to cope with, preventing a knowledgeable internal overview of their activities.

It is therefore critical to establish programs to ensure the quality of the output of these processes. Some of the approaches taken are certification and regular recertification of operators, as mentioned earlier; control laboratory analyses and cleanliness audits for metal joining, plating, and painting processes; test specimen analyses for metal joining and heat treatment; and the use of penetrameters, standards, and bath control for certain nondestructive tests. The use of statistical control charts on key aspects of these processes, especially the operator-controlled ones, is often very informative. Records of use of items that deteriorate, such as certain ultrasonic transducers, can be used to minimize failures through timely replacement.

To accomplish all this effectively when the user of the process does not have the expertise to handle all the technology involved often requires obtaining help from outside. Even when there is good internal capability, the equipment manufacturer or technical service groups often can make useful suggestions and should be consulted.

Some organizations ignore Special Process Control just because they do not understand the technology and the risks associated with the processes. Since these processes often contribute critically to the quality, reliability, and safety of the product and to ultimate customer satisfaction with it, it is important that appropriate control programs be established and followed.

4.4 Quality Check Stations

Whether dealing with an established production process or preparing for the installation of a new one, it is imperative that proper consideration be given to the logistics of required inspection and test, just as it is necessary to plan the production process flow. To do this effectively requires an understanding of the process and its capabilities (see 6.5 Machine and Process Capability), of the relative importance of the characteristics of the product (see 6.1 Classification of Characteristics), and of the relationship between product and process (see 2.6 Design Responsibility for Manufacturability and 2.9 Product Design Review).

With this background, proper decisions can be made on what receiving, in-process, and final inspections and tests should be performed and who should perform them (either Production or the Quality function, either operators or inspectors); at what points in production and how frequently; what equipment and facilities are required; what handling and protection are involved, with associated storage space for rejected material and for that awaiting evaluation; and the purpose of the check. In determining purpose, the check can be of a release or clearance type, such as for approval of purchased material or the setup of a process, approval of the first item(s) produced from a setup, or piece or lot acceptance; or the check can be of a control type, with "roving" or "patrol" inspection or audit of a process, or product audit after "end-of-line" evaluation.

Whatever decisions are made, certain results of these conclusions should be borne in mind. For most organizations, appraisal costs are high. When problem-prevention efforts are ineffective, appraisal is all that protects the producer from unknowingly flooding his assembly operations or his customers with nonconforming product. Even when effective prevention investment is made, appraisal will remain high for a while until management is satisfied that sufficient control has been achieved. Also, customers and regulatory agencies may have assurance demands that require extensive appraisal effort, regardless of the demonstrated quality of the product. Therefore, proper resource management requires careful planning and decision making in this area. (For a discussion of the data generation, processing, and reporting aspects of this inspection work, see 5.3 Received and Produced Quality Data Reporting.)

Economies result from early detection at receiving inspection and at the start of production of nonconformities. This is implicit in the setup and first piece clearance inspections, but it also influences operator self-inspection and automatic in-process inspection decisions. Roving inspection and audit activities, when properly oriented, also contribute to this consideration, as does product audit after acceptance, in warning of nonconforming product ready to ship to the next department or to the customer.

Once all the decisions are made, the necessary equipment should be made available (see 7.5 Quality Measurement and Control Equipment Applications); facilities such as utilities, space, benches, and racks should be provided; and detailed instructions should, in all except the simplest cases, be prepared and provided to the inspectors and testers. The instructions should be presented in the most effective format and should be compatible from area to area, permitting personnel transfer with minimum retraining of the planners as well as the workers.

Visual aids, such as photographs, exploded-view or isometric drawings, cathode ray tube presentations, and physical samples are often used to supplement or replace written instructions (see 1.2 Quality Standards). The minimum level of inspection and test operating instruction is a marked-up drawing, which is woefully inadequate for the purpose and consequently uneconomical, in most situations. Up-to-date instructions (see Appendix B) must be readily available at the work place or with the auditor, roving inspector, or other person doing the evaluation for maximum utility and for ease of reference by anyone concerned with the work being done. A key consideration in optimizing the overall appraisal effort is assuring that the inspection and test plans and instructions are compatible with each other. This implies that methods and standards for in-process acceptance will not have the potential for passing product that violates final acceptance requirements. Controls to ensure this involve coordinated planning for suppliers, the Quality organization, Production, and Field Service. (See 3.3 Supplier Measurement Compatibility, and 9.4 Field and Factory Standards Coordination.)

The control of emergency situations from a planning and operating standpoint is covered in 4.9 Contingency Planning.

When Quality Check Stations are properly established and operated, necessary appraisal will be performed so as to minimize production of nonconforming items, with its attendant costs, and to maximize protection of the customer from the receipt of such nonconforming product.

4.5 Packaging and Packing Control

The product and its associated packaging and packing materials must be designed compatibly with each other. This requirement is vital when the package will be visible to the customer at the point of sale, but it is usually desirable for any product. The practice is most cost-effective when product and package design take place concurrently and with the involved processes. Therefore, the packaging design function is identified as a participant in 2.9 Product Design Review.

Packaging must be able to support the product and to protect it effectively from shipment and storage damage, including that from contamination by outside agents or by the packaging itself, or from conditions that might cause deterioration. But packaging if properly designed may also be usable by the customer for other than the original purpose, and it may also be made profitably returnable and reusable.

Careful handling and storage of packaging and packing materials before use, to keep them clean and to prevent deterioration, is important because of the potential that packaging has for causing or permitting damage to the product. Stocks should be used on a first-in, first-out basis, just as production material is used.

Actual application of the material should be subjected to process control techniques during the packing, labeling, and shipping operations to prevent damage to the product and to its container and to ensure proper shipment of the correct product. Used containers should be properly cleaned before each reuse.

Packaging and Packing Control will result in increased customer satisfaction and reduced costs for those organizations that apply the concepts imaginatively.

4.6 Production Material Handling and Control

Since, for most manufacturers, direct production material constitutes a significant, and sometimes the most significant, portion of manufacturing costs, considerable effort is usually expended to control the size of the inventory and to turn it over rapidly. Attempts are also made, where shelf-life problems exist, to process the inventory on a first-in, first-out basis.

One of the relatively recent concepts to be applied to the problem of massive material stock and in-process inventory is that of Just-In-Time (JIT)—the receipt of material from a supplier directly at the point of production just when needed. The effect of this is to eliminate receiving inspection of those items, to minimize delivered quantities and thus associated storage, and to reduce inventory costs.

To accomplish these results without causing shortage disasters in their production operations, users find that they must receive only perfectly good material from their suppliers. As a result, organizations adopting JIT to any degree are forced to enter into a new way of buying items (no more purchase quantities driven by Economic Order Quantity precepts, for example) and of dealing with their suppliers (gone is the adversarial approach). In essence, at least in this area, they are forced to apply the Quality System.

Indirect material is often less efficiently controlled, despite its frequent susceptibility to damage or deterioration. And many aspects of the storage and handling of both types of material can affect our ability to ensure personnel safety in handling material on receipt and in process and to produce fully satisfactory product with negligible waste. It is thus to our economic advantage to plan for and execute a quality-oriented approach to handling and control of the material.

Although some of the considerations involved should be obvious, others may be quite subtle. Each organization should establish its own list of items for control, but an example of such a list might be informative:

Receipt and Inspection of Purchased Material

1. Unloading and handling damage potential and personnel safety hazard should be minimized through proper packaging and shipment by the

supplier, availability of correct handling equipment and qualified receiving personnel, sufficient laydown and temporary storage facilities, and required protection against the environment and traffic.

2. The processing of material to any required incoming inspection should be expedited. If samples are drawn from containers in receiving temporary storage, the opened containers should be segregated, have their status identified, and have their contents protected as required, pending disposition of the shipment and return of the sample.

3. Material subject to incoming inspection sometimes must be removed from the receiving temporary storage area and placed in a special location awaiting the results of such inspection and test. Handling and location should be free of potential damage and contamination, with adequate space and facilities in the storage area.

4. The incoming inspection function must be organized to process the material efficiently and rapidly. When lots are found nonconforming, disposition must also be rapid, with appropriate identification and protection of the waiting material meanwhile. Dispositions of rework, repair, scrap, or return to supplier must be expeditiously carried out, removing the waiting material from the storage area rapidly and thoroughly, with due care to prevent its appearance in production stores.

5. When material is returned to a supplier or sent out for rework or repair, it must be carefully packaged to prevent damage. Some electronic parts are highly susceptible to damage from improper handling and packaging.

Production Operations

1. Material released by incoming inspection or upon receipt from the supplier as not requiring such inspection should be properly identified and go rapidly and without damage to production stores or stockrooms, as required. Such store areas and stockrooms must have the necessary facilities, equipment, environmental protection, skilled personnel, and operating practices to ensure nonconformity-free handling, personnel safety, storage, processing, and issue of the material. The stock controls must include the ability to locate and remove superseded material or material found nonconforming by functions other than incoming inspection.

2. Material in process of manufacture must also be identified, be capable of being located and removed, and be protected from handling damage, contamination, and deterioration. The questions of identification and loca-

tion capability raised here and in 1 above, are addressed in 1.9 Material Tracking.

3. Any off-line inspections and tests require planning for safe handling, for the time needed, and (along with in-line appraisals) for the necessary facilities, equipment, and environmental protection to accommodate the disposition time associated with product nonconformity.

Postproduction Considerations

1. Completed items, whether parts, components, subassemblies, or finished product, involve the same concerns as those expressed in Production Operations—as do packing and packaging materials, manuals, labels, and so on. The accessibility of packed product for audit inspections must also be planned for.

2. When the product is transported to warehouses or shipping facilities, warehoused, and delivered to dealers or customers, the prevention of handling damage, contamination, and deterioration is still important. Therefore, packing materials and techniques, selected modes of shipment, and storage facilities must all be developed to minimize quality problems, while at the same time supporting each other and marketing and economic needs as well.

3. The mode of transportation and even the route can materially affect packing decisions. Several years ago there was an effort at evaluating handling damage in Parcel Post shipments. Accelerometers were placed in six identical packages and shipped to selected destinations over different routes. One package arrived very damaged—even the accelerometer was broken. That package had gone through the then-infamous Chicago Post Office, which used a "waterfall" drop to negotiate the different levels of the building. The other routes yielded varying results. Package design and shipment method decisions were substantially influenced by this test.

4. During installation and service of the product, its protection from damage and contamination must also be forwarded, with necessary requirements for equipment and cleanliness specified and enforced.

One of the points emphasized in the foregoing discussion is timeliness, the rapid handling of purchased material and of all nonconforming items. This is important because delay often removes the best of our available options when nonconforming material is found. From the quality and overall economy viewpoints, that best choice is normally to have nonconforming material replaced or reworked. When we delay discovery or disposition of nonconforming material, our own production demands often force us to make less satisfactory dispositions. Therefore, our program for Production Material Handling and Control should be oriented toward rapid processing of material. Since this is wholly compatible with the objectives of inventory control and turnover, it is remarkable how few such programs recognize its relevance.

Inclusion of quality considerations in the plan for handling and controlling material will make it operate more efficiently and economically, with significantly reduced quality difficulties associated with material.

4.7 Manufacturing Quality Plan

Once established in an organization, the patterns of inspection and test are typically taken for granted. When a new product is to be introduced, Quality function's "business as usual" is frequently just expanded to accommodate it, often with little consideration of the potential efficiencies associated with a quality-control approach tailored to the new product. This element promotes the development and application of a tailored manufacturing plan and its associated Manufacturing Quality Plan for each new or changed product or, at least, for each new or changed product line.

By tailoring the plans together, particularly if that is done at an early stage in the product development cycle, useful adjustments can be obtained in the design of both the product and its companion processes. This sometimes can even produce a product-process "marriage" of such character and uniformly high quality output that the separate inspection and test portions of the Manufacturing Quality Plan become negligible in size and cost.

What the element addresses, then, is the documentation of the complete scope of the effort to optimize the production, inspection, and test department efforts to guarantee manufacture of a satisfactory product. This includes those steps involving making improvements in the capability of the process even after it has proven to be better than the specifications call for—"continuous quality improvement" or "continuous process variability reduction" are terms used in this context. The program starts with purchased material, covers process control and final product approval, ends with product installation or erection, and includes all other quality-related activities in between.

Parts of the Manufacturing Quality Plan are developed in other elements of the Quality System. The purpose of this element is to combine all the other activities into a single framework by reference, to add those additional activities that are not covered elsewhere in the Quality System or that are peculiar to the product under consideration, and to balance all of this to produce a workable and economical result that everyone involved can understand and accept.

As part of the balancing effort, it becomes useful to compare the cost of inspecting for a particular nonconformity against the costs of not finding it, including the impact in the field. When the probability of occurrence of the nonconformity is very low and the cost of finding it exceeds the costs of not finding it, most managements would decide to forego such inspection. This condition virtually always pertains when the process variability has been reduced to a tiny portion of the tolerance spread and the process average is nearly centered.

Another consideration in structuring the Manufacturing Quality Plan is to ensure that internal inspections and tests are correlated with field practice and experience. While it is often necessary to inspect or test for characteristics not currently found to be nonconforming in the field, it is always required that those characteristics causing field problems be checked for in-house, and typically at final inspection, test, and audit.

The Manufacturing Quality Plan must include not only the product-process control questions, inspection and test, correlation with the field, and customer operation; it must also be concerned with timeliness of action and reaction, production volume and shipment schedules, personnel safety, and flexibility to accept new and substitute materials and changes to all of them. This means that the requirements for resources—people, their skills, facilities, equipment, and maintenance—must be established with contingency plans for unanticipated changes. To be effective, the Plan must satisfy, and be approved by, the management of all involved company functions. Since it requires constructive reaction to changes, it must provide a mechanism for notification to the planners of impending new materials, new processes, and new equipments as early as practicable.

The formal Manufacturing Quality Plan, properly accomplished, produces optimum resource application in quality-related matters associated with suppliers, production, and field installation.

4.8 Quality Work-Force Determination

It has been customary in many industries for the size of the Quality work force, at least that portion assigned to inspection and test activities, to be controlled as a long-standing fixed percentage of the direct labor production work force. Although this practice simplifies budgeting and personnel acquisition programs, it has nothing else to justify it. With increases in automation and a consequent reduction in productive direct labor, appraisal effort has been automatically reduced, even though a concurrent increase in product complexity might well have demanded increased effort to protect the company and the customer.

Similarly, automation of inspection and test along with production operations might well have justified a disproportionately larger reduction in appraisal manpower than in production. What should be apparent is that control by ratio in this area is of no more utility than in any other—essentially none.

The requirement is for careful study of the real needs for personnel to perform quality-related tasks. Industrial engineering techniques of time and motion study, method analysis, and the establishment of engineered time standards are just as applicable to repetitive inspection and test activities as they are to production-line jobs. Even nonrepetitive complex work can be synthesized and have useful standards developed. With the job requirements properly defined in advance, as described in 3.1 Quality Information Package for Suppliers, and 4.4 Quality Check Stations, these time standards then provide the logical basis for determining the skills needed by and the size of the appraisal work force. Decisions to automate inspection and test are also supportable by such studies.

Whether the personnel report to Production or to the Quality organization, whether they are direct or indirect labor, or whether what we are dealing with is production labor appraising their own work, is immaterial. The approaches described are applicable in any case; the size and composition of the Quality-related work force are properly determined by the job to be done. Once agreement is reached on the magnitude of that job, often at least partly dictated by customers or the state of the processes, the decisions on work force size can be reached objectively.

With the standards established, appraisal and related training costs can be accurately forecast and personnel performance can be effectively controlled, as described in 8.12 Quality System Personnel Performance. The introduction of new or changed products, processes, or quality plans will be reflected by appropriate changes in the quality work force and costs, regardless of any direct labor changes involved.

By making the Quality Work Force Determination as proposed by this element, the organization can optimize its quality costs and obtain consistent inspection and test coverage throughout its operations.

4.9 Contingency Planning

Regardless of how thoroughly plans have been made (see 4.4 Quality Check Stations and 4.7 Manufacturing Quality Plan) to ensure the control of purchased or produced material quality, situations will arise when emergency or short-term modification of or additions to those plans will be needed. We may have to accept a different, but usable, material for a few lots, often involving a "better" material than specified; we may have to obtain an unplanned-for item on a crash basis; we may need to evaluate a new or revised machine or process; or a key piece of processing equipment may unexpectedly break down, requiring the use of a different process until repairs can be made. At such times, we have to vary from our standard quality plan briefly or for a few lots.

A different type of need for Contingency Planning occurs when one has determined to make a permanent change but must act immediately, the permanent change mechanism requiring more time for paper work and approval processing than can be tolerated. The plan may have been found unworkable, perhaps under field conditions. Or it might be discovered that a critical test has been omitted in a customer-approved inspection and test plan, requiring formal revision and approval of the plan. Meanwhile, a temporary change to the plan must be made, pending its formal reissue.

All of these circumstances, and others that might arise in some industries, dictate a standard method for responding to the need with an instant change. The Quality System requirements, to avoid anarchy, are that such "informal" changes must be carefully controlled for scope, duration, utility, and responsibility. Spelled out in the procedure must be the types of conditions that may have Contingency Planning applied; the time limits, number of lots, or activity completion criteria that call for management involvement or cancel the temporary plan irrespective of its replacement by a permanent plan; the assurance required for the capability of the substitute process; the provision for controlled extension of the temporary plan; the legibility, effectivity criteria, documentation requirements, and minimum information content of the change; and the specification of personnel and functions authorized to apply and to approve the temporary change method. And they must be adhered to in satisfying the organization's needs for control.

The Material Tracking (1.9) program may require that items or lots processed under temporary plans must be so identified or registered. Other records of the use of the Contingency Planning procedure may be required. These requirements should be attended to in the procedure.

This program provides needed flexibility in the quality planning process, but it does so in a manner that ensures the proper level of management control of such flexibility.

4.10 Inspection and Test Performance

In every manufacturing enterprise and in many service ones, it is necessary to conduct inspection or test of some or all of the details of purchased material, equipment setups, process parameters, material in process, finished product, packaging and packing, installed product in the field, and field service activities. These appraisals are conducted to assure management of the adequacy of preventive measures taken to obtain satisfactory quality results, to prevent work being invested in already nonconforming material, and to minimize the likelihood of customer dissatisfaction with the product as received, installed, or serviced. In addition, properly planned and conducted inspections and tests produce quantities of data that can be used to identify the causes of nonconformities and to provide timely guidance for their correction and the effects of corrective actions taken.

Since much money is often invested in this appraisal work, the inspection and test functions must be properly engineered for optimum efficiency and effectiveness, with the provision of the correct numbers of fully trained personnel to install, maintain, calibrate, and operate the right equipment for the job. They must then be managed to perform in the most effective and expeditious manner practicable. (See 4.4 Quality Check Stations, 4.8 Quality Work Force Determination, 5.4 Appraisal Activity Evaluation Data, 8.4 Inspector and Tester Training, 8.11 Qualification Standards, and all the elements of Subsystem 7, Quality Measurement and Control Equipment.) This need for efficient engineering and management exists whether the appraisal function is being performed by automatic machinery or by personnel of the Quality organization, by Manufacturing, by Installation or Service, or even by those of the engineering laboratories.

Whoever is performing the inspection or test or controlling the automatic inspection and test equipment must generate and report the resultant quality data of the correct type, amount, and validity. Planning for this is covered in 5.1 Quality Data Planning, and its actual generation is described in 5.3 Received and Produced Quality Reporting and 5.5 Field Quality Data Reporting Control. Proper conduct of this aspect of inspection and test leads to the expeditious release of "good" material or to the timely application of the Corrective Action Program (1.12) and ultimately to the elimination of the causes for nonconformance, reducing the need for the routine, expensive types of inspection and test.

4.11 Inspection and Test Support

Laboratories and inspection and test organizations are typically equipped and staffed to perform the routine analyses and other appraisals needed to support production. Sometimes this capability is expanded to support research, development engineering, and field failure analysis activities as well. Even in the latter situation, however, the occasional requirements for such services have not been planned for specifically, and often these unscheduled service requests are for vital information to support design decisions, machine repair, product liability defense efforts, or competitors' product evaluations. Therefore, the organization must be able to provide such service as needed.

Any appraisal organization tends to have a varying workload while performing its primary function. Nevertheless, it often is desirable, particularly in laboratories, to consider providing excess capacity for covering the unscheduled demands described in this element. The excess capacity usually can then be profitably marketed in the surrounding industrial community. Such an approach can support special equipment and technician talents for occasional internal use while providing all members of the organization with a broader range of experience than they would otherwise obtain.

One administrative problem frequently encountered is that of maintaining proper control over special appraisal requests. Often an engineer will ask for a special study to be run on a part to provide data for a design or disposition decision. If the request is somewhat vague, the appraisal agency may continue to perform these extra measurements indefinitely. To control this sort of problem, many organizations have adopted a request form that calls for specific information covering the areas of ambiguity and that also may be used to charge out these services, if desired.

Another problem arises when there is a lack of control: People ask for special studies that are not really needed. The approval process associated with the handling of the request form can reduce this problem. If the program calls for timekeeping, management decisions about the benefits and scope of the service can be reached objectively, based on program usage and personnel involvement trends. Such studies can also identify needed additional or different equipment and skills.

In determining what techniques or equipment are to be used to make the requested appraisal, sufficient planning is required to balance available skills and facilities with the needs of the requester. Similar planning should determine the minimum amount of data needed and provided to satisfy the requester's requirements.

When the Inspection and Test Support program is properly operated, the needs of the organization for special appraisal services can be satisfied, often at an actual monetary profit.

4.12 Housekeeping and Plant Safety

"Cleanliness is next to Godliness." While this slogan was not developed in an industrial context, its applicability to achieving quality in a product is unquestionable. Certainly the valid concern evidenced by strict government cleanliness regulations in the pharmaceutical industry, by a similar recognition of the necessity for "clean room" environments for the manufacture of aerospace components and nuclear reactors, and by food and beverage controls illustrates a significant understanding of the potential contamination hazard to these products and to their users.

Levels of nonconformance not high enough to be deemed catastrophic are often ignored when their causes are not immediately obvious; many times these causes are directly related to contamination. While the broad question of process contamination, particularly from the outside, is covered under the subject of 6.4 Environmental Impacts, the source of many mysterious nonconformities is often the lack of internal cleanliness—in a word, Housekeeping.

Manufacturers not in the cleanliness-related industries are often unaware of the impact of unsatisfactory Housekeeping on their own products and operations: Close-fitting parts are often damaged or subjected to rapid wear-out because of foreign material between them or in the lubricant; blow holes occur in castings, weldments, and solder joints because of volatile inclusions; fumes produce discoloration or other undesirable chemical reactions during processing; dust, vapors, and particulates cause explosions in enclosed areas.

The reduction in internal failure costs, the improvements in employee morale, and material usage efficiency are benefits of a thorough dedication to cleanliness in the factory, and they far more than offset the costs of a positive Housekeeping program. Defined tasks, effective supervision, and participative programs, combined with well-designed audits, will produce the desired results with minimum cost. The largest single cost saving for which I was ever responsible, more than $600,000 a year, was the elimination of a scrap-producing nonconformity caused by a lack of cleanliness in the factory (unsuspected by those responsible for the process or the product). The program to eliminate the nonconformity cost about $500 a year to administer.

There is often a direct relationship between Housekeeping and Plant

Safety. Every accident or near accident should be treated as a quality failure. Its root cause should be determined and eliminated and become part of the 1.12 Corrective Action Program when this does not occur. Safety, morale, and other employee-satisfaction indicators should be built into the quality-related data bases of 5.1 Quality Data Planning. Other safety considerations are covered in 2.5 Safety Engineering.

4.13 Yield Control Program

In the context of this element, "yield" may be viewed as the fraction of satisfactory results from the total output of a process. In that sense it can be used for any activity whose results are measured in some consistent fashion. Manufacturing organizations use the term to describe the percentage of acceptable parts from an operation or a process. Some of them consider only "first-time" acceptance; others measure the total yield after rework or repair.

When we talk about a Yield Control Program in the Quality System, we are concerned with more than just measuring and reporting yields. We are concerned with identifying where yields of less than 100 percent could have a damaging effect on operating costs or, more significantly, could affect product performance or reliability as the result of defect "leakage" through our inspection and test screens. Therefore, we must set minimum limits, frequently by part and sometimes even by characteristic (see 6.1 Classification of Characteristics), on the permissible yields of key processes that can significantly affect economy, reliability, or performance.

To be able to set such limits effectively requires knowledge not only of the capabilities of our processes but also of the contributions made by the parts and their characteristics to product performance and reliability. Often certain statistical and forecasting techniques are necessary to permit valid selection of those conditions needing control and to determine its proper extent. Such considerations affect decisions on process equipment purchases and on incoming and process sampling schemes.

Another important consideration is to view all of these calculations and projections from the system standpoint—for example, correlating the purchased item decisions with those of the processes, to be mutually supportive. In addition, purchased item and finished goods yield controls may have to be applied to such areas as preservation, packaging, and packing in order to maintain the produced quality level through shipment and for a variable period in storage.

Once the required yield limits have been set, it is necessary to see that they are met or bettered. This requires a record and exception and trend reporting program that provides rapid indication of limit violations and impending violations to the personnel charged with taking and ensuring effective corrective action. (See 5.3 Received and Produced Quality Data

Reporting.) Such a program also provides the tracking mechanism for continuous quality improvement efforts.

In the initial stages of corrective action, an important step is to verify that the indicated poor yield is real. Ineffective Inspector and Tester Training (8.4), outdated or nonexistent Quality Standards (1.2), or failures of the Equipment Calibration and Maintenance (7.6) program all may result in erroneous identification of nonconformities. The latter condition becomes evident when the poor yields are investigated and, quite often, when field quality data are validated under the provisions of 5.5 Field Quality Data Reporting Control. Of course, the potential also exists for these same activities to result in incorrect "good" yield identifications.

Once the yield exceptions have been validated, corrective action should be undertaken immediately, utilizing the Corrective Action Program (1.12), if necessary. The effort may involve a relatively minor process adjustment or product design change or could require the initiation of a Quality Level Improvement (6.7) project. Whatever is involved, the responsible supplier or company personnel are expected to expend or to obtain the needed effort to achieve the yield goals expeditiously.

Although the exception reports of this element and those of the Quality Cost Program (1.7) will sometimes highlight the same problem, often they will not. One major reason is that quality costs associated with later operation or field problems frequently cannot be traced to their real source, but the yield exceptions get to the problems where they occur. Another major reason is that the "fix" for a yield problem is often to add a standard operation, such as tumbling to remove burrs, that is not then identified as a quality cost. If we were to examine the yields of the process that produced the burrs, however, it might be obvious that there is a quality problem that might be preventable and not require the extra manufacturing operation. Thus, the Yield Control Program provides a vital control mechanism, one of the six key control devices of the Quality System, for assigning resources effectively to obtain required operating economies, product performance, continuous quality improvement, and product reliability.

Subsystem 5

Quality Data Programs

Subsystem 5 is intended to organize the bulk of the Quality System's data requirements into the most manageable, usable, and economical form. The elements are arranged in the order for most effective achievement of this objective, supplemented by a few in other subsystems.

Ever since the earliest days of modern quality control, it has been recognized that records and data are the lifeblood of the program. Statistical techniques involve the structured gathering and analysis of data. Extensive record keeping is necessary to assure quality of the product shipped to the customer, to fulfill many contractual and regulatory requirements, and for adequate defense in product liability cases. The resultant information is also needed for comparison with competitors to ensure that the organization is properly responsive to needed improvements in quality and other determinants of customer satisfaction. When we examine the aggregate need for quality-related information, it is mind-boggling. In most organizations, however, the subject has never been looked at in its entirety.

The need for this mass of data did not arise suddenly and all at once. Instead, small chunks of it were developed in response to different stimuli at different times. Incoming inspection data needs arose because of the requirement for effective communication about quality problems with errant suppliers through a dubious purchasing and sales community. Process quality data needs arose in response to organizational growth and departmentalization; later on, the process quality engineer added requirements for data to aid in statistical process and machine capability studies, demonstration of the state of control of the processes, and troubleshooting. Final product inspection and test data were needed to support product release-for-shipment decisions and then to answer questions from customers about specific item quality.

Next, the development engineers began to recognize the need for detailed knowledge about the components of their test models. Later on it became apparent that the organization needed to have factual data about the product's performance and failure patterns in actual use. Ultimately, we began to concern ourselves with the costs of all this quality effort, and additional demands for data arose in many previously unidentified areas to provide answers to the question of why quality costs were high.

Once these data programs were developed (actually, in many cases they just happened), they tended to remain intact even as needs changed. Inspection and test departments became repositories for vast quantities of data that were rarely, if ever, used for any purpose other than the original item disposition—for which purpose more information was often generated than was actually necessary. At the same time, the information needs of the other purposes were often not anticipated and were not adequately served by the data generated. It has been said that inspection departments were the primary contributors to the success of file-cabinet manufacturers.

With the advent of the computer, the opportunity arose to generate, analyze, store, and report quality data almost painlessly—in just the quantity that properly served the needs of all users. Unfortunately, many organizations merely transferred en masse the contents of their file cabinets to the computer without approaching the data question from a "system" viewpoint. It is necessary to identify exactly what data are required by all users and then to develop the most efficient way of providing them.

In addition, computer programs have caused a significant increase in the awareness of users and customers that many and critical quality problems and failures were hidden in their design and operation. As a result, "software" quality assurance is now of great and growing concern. This becomes of even greater importance when we recognize that information networks are proliferating in number and becoming global in scope. See 5.6 Software Development, 5.7 Software Testing, and 5.8 Software Operation and Maintenance.

5.1 Quality Data Planning

The Quality System requires that all the needs for quality, reliability, morale, turnover, safety, and environmental data within the System be identified; that the best methods and sources for obtaining just those data on a timely basis be located; that the time demands on operators, inspectors, and testers for data inputs be minimized; and that provision for rapid retrieval and analysis be made, both routine and on demand. It further insists that records retention cycles be set compatible with the needs for 1.8 Product Liability Prevention, and that exception, trend, summary, and status reports be provided as appropriate at minimum cost and with requirements for root-cause determination and needed corrective action in consistent formats to provide familiarity and minimize interpretation problems. This overall process can be described as records engineering. It correlates with 5.2 Forms Control. Although, when viewed in its entirety, the data requirement of the Quality System is so mammoth that it ordinarily cannot be satisfied except by the use of the computer, small companies and certain remote entities of large ones may find a manual approach necessary.

In one application of good records engineering, a manual incoming inspection data program requiring about 5 percent of the involved personnel's time was developed to replace one that required an average of 50 percent of their time. Interestingly enough, the utility of the data was greatly increased simultaneously. For all practical purposes, the effective size of the department was doubled without any increase in head count. This was accomplished by consolidating into one report eleven slightly different forms that had to be filled out to notify eleven different persons of the receipt of each defective lot of material, by preprinting basic information on the single form so the inspector had to enter only the minimum of specifics, and by having the report generated mechanically by the preprinting equipment, a simple electronic data processing application.

What can be done with the computer is even more striking. One sizeable facility has placed computer-connected inspection and test equipment and input terminals in the production areas. Quality data are entered both automatically and manually and, when limit criteria are exceeded, the computer alerts the responsible supervisor. It then provides specifics about the problem with rank-ordered suggestions for corrective action, based on its files of successful corrections previously applied for the same or similar conditions. The supervisor resolves the problem and informs the computer of how the resolution was accomplished. The computer adds the information to its data base, judges the results from later inspection inputs, and then updates its solution-ranking file. In the process of investigation and correction, the computer assists with specialized data analyses, including graphic displays. The net result has been an 80-percent reduction in the process inspection force and the virtual elimination of the continued production of any important defect in quantities of more than two. All of this resulted in a significant reduction in both Appraisal and Internal Failure costs.

One further consideration is the necessity of accumulating data for purposes of comparison. We need to know not only what we have done and are doing, but what our competitors have accomplished and what has been happening in other comparable organizations around the world. This enables us to engage in Quality improvement efforts oriented to those situations where we are weak and those where we are potentially able to equal or surpass our competition. These comparative data points are often referred to as "benchmarks," and they should cover every part of the business—employees, products, processes, service, and non-product-related operations.

Properly engineered and operated, the Quality Data Planning program can be economical and effective in satisfying all the various demands upon it described above.

5.2 Forms Control

Despite the advent of the computer, with input terminals available to the inspector or tester or automated appraisal equipment feeding data into the data base directly, an enormous amount of quality data is still recorded and processed manually. In addition, it is often necessary to produce many kinds of quality-related reports that must be sent to customers, suppliers, other intracompany factories, or to headquarters. Format engineering or, more broadly, Forms Control is oriented toward achieving the necessary transfer of information as rapidly and economically as practicable.

Multipurpose forms, which serve the needs of two or more functions receiving data from a single source, reduce not only the number of forms that an inspector must initiate as the result of completing an assignment but the amount of time and money involved as well. I once found an inspector who had to complete thirteen different forms, all containing much the same information, to dispose of a defective lot of certain types of purchased materials. Careful engineering reduced this work load to three forms, which satisfied all real requirements with about 20 percent of the inspector's previous time investment and no measurable loss of efficiency by the recipients of the reports. The concept involved is "standardization with zero duplication."

Zero duplication means that the person originating or processing any specific item of data should do so just once. Standardization implies that a given type of data will be found in the same area of all forms that contain it. This practice habituates initiators and users alike to the data pattern, producing rapid completion and processing of the forms. Transcription into computer files or other mechanized records is thereby accelerated and made less prone to error. Appraisal personnel, clerks, and data users transferred to other parts of the organization experience fewer delays and make fewer mistakes in adjusting to the new environment insofar as the data requirements are concerned.

Achieving really effective Forms Control implies careful forms engineering and centralized control of report formats, form numbers, and ordering. Although modern duplicating equipment precludes this type of control over single-copy forms, effective control of computer outputs, multicopy forms, and hard-copy tags is still obtainable by the methods described.

Economies from the Forms Control program can be realized through larger printing orders of fewer forms, reduced errors, and increased early effectiveness of transferred employees.

5.3 Received and Produced Quality Data Reporting

The need for determining which quality data must be generated to satisfy the needs of all users is covered in 5.1 Quality Data Planning. Looming large among the voluminous quality data requirements, particularly for manufacturers, is the demand for data associated with inspection and test. This encompasses data about purchased items, about machines and processes and of products-in-process, and about finished product, including after packaging—and shows trends in these data. These demands are the sphere of the Received and Produced Quality Data Reporting element.

In all cases, we want the data to be generated, processed, and reported as rapidly as possible to permit necessary corrective action with the smallest practicable production of nonconforming items. Such effort minimizes the economic pressures to accept nonconforming material and restores the system to a preventive mode of operation with the least delay. We also want the data program to be operated at minimum cost while satisfying everyone's needs for quality information, as emphasized in 5.1 Quality Data Planning. This provides change-directing information to Development and Manufacturing Engineering, to suppliers, to production supervision, and to others involved in product and process planning and operation.

Quality data include examples of demonstrated material conditions as well as test and inspection numbers and descriptions. This means that, where useful, samples of items demonstrating achieved quality should be carefully identified and preserved for comparison with later production. For example, companies have been able to resolve disputes with suppliers by displaying material previously received from such suppliers to refute claims that "they've always been like that." This type of program, particularly valuable for dealing with appearance and other surface imperfections, pertains to internally produced as well as to purchased items.

Incoming Appraisal Data

Inspection and test data associated with received material include those for development engineering application as well as those for production, spares, and service usage. They also include internal and external laboratory analyses of the purchased material and data submitted by the supplier or by source inspection activities. Much of these data will be used in 3.10 Supplier Rating Plans, as will those generated in-process or later that identify supplier-caused nonconformities passed through incoming inspection.

Data from Appraisal During Production

Process data include those produced by automatic recording or control devices on process parameters such as temperature, pressure, and chemical concentration. They also include the manually generated information covering the same type of pertinent process control conditions, as well as the results of inspection of material handling and transfer equipment, to verify their contribution to good quality.

The data associated with products-in-process cover all inspections and tests performed from process setup through the final manufacturing steps on internally produced parts, components, subassemblies, and assemblies, batches of material, or documents—whatever the product may involve. Some of the well-known process control inspections and tests are associated with techniques such as the following:

1. Setup inspection—Such things as the work instructions, manufacturing and inspection equipment, tools, fixtures, jigs, dies, quality standards, operational settings, alignment, and work piece placement are examined to ensure that the operation is ready to begin and, when it is then released and begun, that the probability is maximum that the first item produced will meet specification.
2. First-piece inspection—The first piece or first few pieces produced after initial or later setup are examined to assure that they meet specification and that the process is producing results properly located within the specification or statistical limits for continued production of acceptable product; the process is then released for full production.
3. Patrol or in-process inspection—The items being made are sampled during production to determine whether the process is in control and can continue to produce or to identify what the effects of deliberate changes may be.

Finished Product Appraisal Data

Finished product inspection and test data reflect the results of inspections and tests performed on the completed preproduction and production items covered by the description under "products-in-process." This includes sampling of packed product, with specific attention to transportation tests, safety tests, reliability tests, and so on. Since final product inspection and test serve to verify the adequacy of all the appraisal work that preceded them, as well as providing the basis for product release for shipment and any required reporting or certification to the customer, it is necessary to make sure that the inspection and test equipment and standards used throughout are compatible and correlated with each other. It has often proved to be the lack of such agreement that has caused the most difficult and persistent problems within the product acceptance activities.

The data produced by this element serve as inputs to 1.12 Corrective Action Program, 3.9 Nonconforming Material Disposition, and 4.13 Yield Control Program, as well as to numerous other key activities of the Quality System and of other systems. To ensure that such inputs are valid, it is imperative that the inspectors and testers are fully trained and capable, properly equipped, and completely informed on the specifics of their operations. Although it is not intended that these preconditions be met by this element, the managers of the data program must be satisfied that such conditions exist for them to have confidence in the program results. They must also use verification techniques to be able to guarantee the accuracy of the data.

When this quality data program is properly developed and economically operated, it provides the basis for most effective performance of product development, production, and control of suppliers. It tells when to take action and when to avoid unnecessary action, and it measures the results of both situations. It minimizes the continuing production of nonconforming product and helps prevent any such product from being shipped to the customer. And it provides the basis for an ongoing assessment of the effects of continuous quality improvement efforts.

5.4 Appraisal Activity Evaluation Data

Appraisal involves evaluating the conformance to specification of materials, parts, components, processes, subassemblies, assemblies, products, product installation or erection, and product service. Appraisal activities are typically expensive, time-consuming, and subject to doubts as to their validity and utility. We are concerned with how correctly those performing the evaluations (I will call them "inspectors" for convenience) make their decisions, with how long it takes them, and with the costs and net value of their efforts.

This element is concerned solely with the generation, processing, analysis, and reporting of the data necessary to permit the judgments outlined above. These data summaries and reports are to be used as a basis for decisions about inspectors in two elements, 8.9 Personnel Development and 8.12 Quality System Personnel Performance. This element must assure that the data program is adequate for providing the necessary amount and form of information, including trends, to permit valid decisions to be made. In addition, it must be economical to operate, timely in reporting, and compatible with all related data programs and needs of the Quality System.

In the work of Elements 8.9 and 8.12, it is required that standards of performance be set for the timeliness, cost, and accuracy of the work of the inspectors. This having been done, part of the data program involves the comparison of actual results with the established standards and the highlighting of those relatively few occurrences when the standards are violated.

In the sense of *this* element, as well as in that of the others, inspectors are not limited to a specific department or organization. It is the conformance-measuring work with which we are concerned, not the title or organizational affiliation of the worker.

When the data program is planned and operated as described, personnel numbers and necessary training can be properly determined. This gives an organization the ability to make accurate forecasts of resource needs and applications, resulting in appraisal optimization.

5.5 Field Quality Data Reporting Control

The generation of field quality data is provided for under the subject of Field Problem Controls (9.12), and the analysis of the resultant data and its use as the basis for corrective action, improvement programs, and failure prediction is covered in the description of 9.13 Field Problem Handling and 9.14 Customer Satisfaction Measurement. In all these activities the user of the results must be concerned with the validity, timeliness, and cost of the data program. In addition, the user needs to have the field data compatible with his other quality data programs, for maximum ease of problem investigation and for proper determination of the scope of corrective action. This is accomplished through the use of common defect codes, location descriptions, equipment sensitivities, causes of gains and losses of customers and market, and so on. To satisfy these requirements and to deal most effectively with the very complex field data question, it is useful to have a separate element, Field Quality Data Reporting Control, devoted solely to these aspects of the Quality System.

The validity of the data can be established essentially in three ways: by comparing the field data with related factory data either on an actual basis or through simulation, by comparing the results from one agency with those received from others, and by careful audit of samples of field practices and data using skilled auditors. In all three of these methods, statistical techniques can be employed profitably to test the results, thus giving a level of confidence to the study findings.

When it is determined that the degree of validity is unsatisfactory, programs to correct the situation should be undertaken. These programs may have to be tailored to suit a specific agency condition, or they may be generally applicable. In any event, they should be compatible with, and may include, the data sampling techniques described in 9.10 Field Problem Predictions, 9.12 Field Problem Controls, and 9.13 Field Problem Handling. Some companies have found that their only recourse was to pay for getting good data.

The question of timeliness of the data again is complex. Product sits for varying lengths of time in parts of the distribution chain, so the producer often has no way of predicting accurately when useful data will be received. Early indication of batch-type problems may go unnoticed because the producer is unaware that only a small part of the batch has been used.

The objective, however, should be to get the data reported as rapidly as practicable. Standards of timeliness should be set and announced to the field. Warranty registration cards are often used to provide a base count for items "in use" and for "aging" reports. Audit is also useful here, since a comparison of service records with report submission times can identify areas of needed improvement against the standard. Some companies employ incentives, such as trading stamps, prizes, useful tools, or other recognition, to encourage proper report submittal, using performance against standards of timeliness, quantity, and quality of reports as the basis for the award.

The economics of field quality data collection and processing are a significant factor in the design of the program. A company must be satisfied that it is "getting its money's worth" in improved field results, especially in increased customer satisfaction, to justify what is usually a rather expensive undertaking. But when the data program is properly designed in line with the sampling concepts referred to in the discussion of validity of the data and is managed as described in various parts of this discussion, the program proves to be cost-effective in virtually every case.

Too often the program is developed only to meet the needs of Field Service or as a marketing device, and not as an integral part of the Quality System oriented to serving all users effectively. When this pitfall is avoided, the field data program can be both economical and a valuable tool for improving design, manufacturing, and service ability to satisfy the customer.

5.6 Software Development

Once the software Product Description (2.2) has been generated, the process of development can begin. This effort has several stages, starting with a view of the entire system (software and hardware) and continuing through the interface between the two into the basic structure of the software, together with its associated logic.

It is important at this stage to incorporate the beginnings of the plan for achieving the requisite quality and reliability of the ultimate product. For purposes of achieving productivity of cost and schedule in development, use, and maintenance of the software, we must design the quality into it instead of relying on final inspection and test to uncover its inadequacies. Consistent with this approach, we should also provide the plan for generation, analysis, and reporting of quality data from every activity from this point on so that we can measure the quality of the development process and the completed product. This will give us the capability of predicting further processing and user experience of fault density and failure rates—the basis for process improvement for future projects, for the reliability of the product, and for long-term marketplace success.

The plan should provide that all data and resulting information be traceable to the source to support corrective action and be subject to rapid feedback to the involved development personnel. All quality measurements should be goal-oriented and quantified, based on the original specifications. When this is not practicable, subjective measures may be used.

We must also concern ourselves with the timeliness and quality of the forthcoming documentation—including diagrams and charts—so that user and maintenance personnel can understand fully what is occurring when they encounter problems with the interim and complete product. Wherever practicable, the documentation should be written in clear language and standardized and computerized for ease of use.

In addition, the plan must address the question of testing the resultant coded instructions during the coding process and as a final product. Since testing of complex system software typically is done in several stages (see 5.7 Software Testing), it is essential that the programmer configure the design to facilitate achievement of these critical quality steps.

Programming itself then takes place using a variety of guidelines, methods, and tools, as well as the above considerations. In any organization it is helpful to set up a library of such items. Select those that appear to be the most useful, flexible, and compatible for the types of software involved, then train all personnel in their use. It then becomes a matter of internal discipline to ensure that the selected approaches are consistently applied. In many cases it is necessary to develop algorithms and databases as part of the programming activity. These, too, should be standardized as much as possible—with appropriate and effective interface mechanisms.

A well-engineered Software Development process can produce an end product that contains a minimum of "bugs." Given that kind of experience, extensive testing requirements can often be reduced to a more manageable and economical level.

5.7 Software Testing

The overall software project plan contains the basis for inspection and testing both during the process of development and as a final product. (See 5.6 Software Development.) Such testing usually is planned to take place at the end of each phase of development. Thus, tests are performed after each software component is completed, for each interface/integration of two or more components, for the complete system, and finally from the user's point of view. Obviously, each attempt by a user to apply the program in his environment provides a further test—the resultant data from which should be fed back to the designers for appropriate action.

During the development process, the quality-control function usually focuses on compliance with applicable standards, satisfaction of customer and other specifications, and consistency of approach. In some cases, the use of statistical process control techniques is possible, but often it may be impracticable because of a lack of comparative data. Only when the effort is focused on the process rather than the product can these techniques be applied successfully. In those cases, the control mechanisms can be used to disclose the patterns of variation of the universal key parameters of all projects.

Certain organizations use the goal-setting practice to a greater extent than others—defining in-process targets as well as final requirements and then testing or reviewing progress toward them. Often, interim inspection is aimed primarily at detecting bugs in the program and then having them removed so that no more work is done on defective product. It is useful to supplement that activity with finding the causes for the bugs and taking steps to eliminate those causes for future projects (see 1.12 Corrective Action Program).

All of the above data are combined with field experience and used as the basis for the determination of quality improvement opportunities. This usage by itself requires that the data be tracked consistently and be traceable back to the test or field environments that produced them.

While Software Testing is virtually inescapable, as programs containing on the order of a million lines of code or more are developed, the possibility of exercising all the potential failure modes becomes vanishingly small. In those cases, alternatives to 100 percent testing must be found. Obviously, control of the design process is the answer of choice, involving a dramatic cultural change in the programming community. Self-checking methods and tools, built into the software, become even more necessary than they are currently.

What is required is a mind-set that, from the top of the organization to the programmer, is committed to finding ways to achieve requisite quality economically in all cases, is dedicated to a continuing quality-improvement strategy, and is oriented toward eliminating the need for inspection at all stages of the process. The stage is then set for effective reuse of proven programs with negligible risk and for reaching complete customer satisfaction with the product upon initial release.

5.8 Software Operation and Maintenance

Once the software product or hardware/software system has been distributed to the customer, we come to the demonstration of the effectiveness of all our efforts to satisfy the customer with it. Unless we have been essentially perfect in performing the programming, sooner or later the customer will find some unsatisfactory conditions in the software as he applies it. Note that I did not say, ". . . perfect in testing the software . . ." (See 8.12 Quality System Personnel Performance for the reason.)

Depending on the circumstances, either customer personnel, a dealer, or we will have placed the product into operation for the customer. At this point the software will undergo checkout in the customer's environment. Assuming no catastrophes occur, we must be prepared to develop and provide updates for some time afterward. Conceptually, at least, such updates should be developed under the same controls as were applied to the original program: responding to the new or changed conditions, developing new or better functions, or improving maintainability. Where pertinent (or contractual), we must provide technical assistance to the customer upon request.

Eventually the product will become obsolete or the producer may find it impractical or improper to maintain it. Under those circumstances, the customer must receive sufficient notification and information to permit him to deal effectively with the problem this termination of service presents. The information to be provided would include alternative service sources, possible contract training for customer personnel, and specifics on replacement products.

The Software Operation and Maintenance element, including final disposition of obsolete product, typically covers the longest time period of any of the phases. Assuming that the top management of the software-providing organization hopes to remain in business, it then becomes extremely important that their concern for quality during this phase be no less than during development, for example.

Initial customer satisfaction with the product (and willingness to purchase another from the same organization) can readily diminish if service during this period is perceived to be ineffective or inadequate. IBM made its initial reputation in the computation field from its top-notch service operations—not from the durability of its equipment!

Subsystem 6

Special Studies

The Special Studies subsystem provides the basis for applying the key programs that, over the years, have been the purview of the quality control engineer. The elements of this subsystem cover activities that involve extensive investigation or experimentation, usually statistically based, or the development on demand, rather than continuingly, of programs differing from normal. Most of the activities entail sophisticated mathematical decision-making techniques, applied by specialists, who develop the studies to the point where they can be readily utilized by nonspecialists.

The elements of this subsystem are arranged alphabetically.

6.1 Classification of Characteristics

The purpose of Classification of Characteristics is to supplement the information contained in hardware or software product, process, preservation, and packaging drawings and specifications with specific additional guidance for the most effective application of resources. It is a device by which the designer communicates to everyone else what they are to give attention to. It is not an evaluation of the actual degree of nonconformance of a characteristic, which is covered in 6.2 Classification of Nonconformities.

Under most circumstances, suppliers, producers, inspection and test planners, installers, and service agencies do not have unlimited resources with which to work. Therefore, they must make choices of where to concentrate their operating and control efforts. It is often thought that tolerance spreads tell this story adequately, but they do not. The same tight tolerance band on a part dimension may be associated in one case with a potential safety hazard, in a second with a performance limitation, in a third with an assembly fit problem, and in a fourth may reflect only established custom that has no application in the current design—all four conditions, perhaps, on different dimensions of the same part.

For the person designing the process to produce this part or developing the quality or installation and service plans for it, the tolerance band on the drawing does not give enough information to enable him to make decisions the way the product designer, who knows the most about the application of the part, would prefer. The same thing is true about process, or intermediate, dimensions and tolerances. Without classification of these characteristics, the manufacturing planner fails to sufficiently inform the manufacturing people and the Quality function planners about the process requirements needing their support.

Manufacturing processes selected for use are never capable of producing all characteristics of any item with equal conformance to tolerances, and inspection and test operations are never performed on all characteristics of the item with equal frequency or with equal discriminating ability. Therefore, since the resource application decisions must be made, it is more economical, effective, and efficient that this decision-making process occur under consistent ground rules. A thorough Classification of Characteristics program provides such a set of rules. However, it must also be understood that *all* information from the designer to the producer must be provided in accordance with a well-planned and consistently applied drafting or documentation standard.

Although the rules must be understood by everyone in the planning and operating flow, how the resulting information is used may well be different. For example, if the designer identifies a characteristic as of the highest priority for user safety reasons, the manufacturing planner might choose to produce that characteristic by a very expensive process, but one that, once adjusted to satisfy the requirements, cannot fail to produce satisfactory items. Such a manufacturing process capability might then show that, despite the critical nature of the characteristic, the inspection and test planner's resources should be applied to this critical characteristic only slightly and be more extensively applied elsewhere. This process of optimization of resource use obviously requires interchange of decision information among planners but, especially where planning groups are large and separate, is dependent on proper initial identification of the relative importance of characteristics.

In applying the classification process, we need to decide both on how many classes we require for proper differentiation and on the decision rule for separation into these classes. Although the number of classes has varied historically from two to six, and even more, the typical value, probably influenced by military documents, is four. From military usage, the most common terminology for these classes is "critical," "major," "minor," and "incidental." For ease of communication, we will use these terms also.

The decision rule, although for everyone's use, is of primary importance to the designer, who is enabled to focus on making the decision without worrying about intangibles and extreme conditions. A typical rule might be as follows:

> Each classification will result from a determination of the effect (during useful life) on safety, performance, reliability, manufacturability, quality, saleability, serviceability, or handling damage of a small deviation from tolerance or specification requirement. A small deviation means one not greater than X percent of the tolerance spread, or equivalent.

Typical values of X in this rule range from 10 to 50 percent, with the latter predominating. Since most designers approach tolerancing from the conservative side, the tolerance limits themselves do not present a useful decision point. This is essentially true, also, of the X = 10 percent and other very small deviations. But after we have conceptually doubled the tolerance spread of a two-sided tolerance with X = 50 percent, the decision usually becomes much easier.

Once the decision rule is developed, we must then establish working definitions of the classes based on the rule and on the realities of our product.

Class definitions might be as follows:

Critical—When a small deviation *will* produce or lead to a substantial safety hazard to anyone or a complete performance failure

Major—When a small deviation *will* produce or lead to some safety hazard; significant performance or reliability reduction; complete loss of further manufacturability; substantial quality, field service, or saleability problems; violation of externally imposed code provisions; or highly probable handling damage

Minor—When a small deviation *may* produce or lead to minimal safety hazard; some performance or reliability reduction; substantial manufacturability problems; limited effect on quality, serviceability, or saleability; or some handling damage

Incidental—When a small deviation *cannot* produce or lead to any safety hazard, performance, reliability, quality, serviceability, saleability, or handling problems but *may* cause minimal manufacturability problems

Classification of Characteristics is applied properly during the first stages of product and process design, is subject to design review, and is updated as the development process proceeds. It is applied to drawings and specifications for purchased material; for internally manufactured parts, components, and assemblies; for "green" or intermediate dimensions; and for packing and packaging materials. It covers nondimensional characteristics also, with decision rules developed in the same spirit and with the same specificity as for dimensional tolerance situations. One-sided tolerance situations are handled in a comparable fashion.

Sometimes the classification information is supplied in a separate document, but it is most informative and effective when it is made a part of the drawings and specifications, usually indicated by a symbol. Symbols should be applied for all cases, although that for the least important class, "incidental", is often omitted in the interests of drafting economics. The hazard, of course, is that an error of omission of a higher-class symbol results in an automatic "incidental" classification for that characteristic. The symbols to be used vary, although standards are being developed that address this question. To me, the simpler, the better, so I prefer:

C = Critical
M = Major
N = Minor
I = Incidental

particularly in specifications. The process classes could use the same letters, perhaps enclosed in a triangle.

It is often best to conduct the classification process in parts. The product designer is responsible for considering only safety, performance, reliability, and saleability. The supplier, internal process, installation, service, and quality planners may upgrade (but not reduce) those classifications when appropriate because of manufacturability, quality, service, or handling damage considerations. This approach also serves those situations when a product is to be built in two or more dissimilar manufacturing facilities.

Sometimes it may be worthwhile to limit the process-related classifications to three categories, omitting the "critical" one. This would be done because these classifications are typically restricted to addressing considerations of process capability, effects on later manufacturing operations, and field service impact.

Whether there are three or four process categories, the definitions are often distinguished by reference to measures of process capability, such as Cpk. A typical approach, then, might have the Critical class cover those conditions where the process capability is known to be out of statistical control and the proper corrective action has not yet been found, or the capability is unknown and unknowable until production actually begins. The Major class might have a Cpk less than 1.00, the Minor class might have a Cpk from 1.00 to 1.33, and anything better than that would be Incidental. In the three-class approach, the above definitions of Critical and Major could be merged.

In addition, the process classification definitions would contain appropriate expressions for the effects of a small deviation on later processing, assembly, and field service. Suppliers would be trained in the Classification of Characteristics discipline and be required to perform the process classification based on their knowledge of their own capabilities. They would then report the results to the customer representative identified in Purchasing's request for quote.

Whether the process classification is provided by supplier(s), internal manufacturing, or assembly (or a combination of any or all of them), the results must then be examined in light of the corresponding product classifications. One excellent way to do this is to compare the actual situation with a generalized matrix of the possible classification combinations. In the four-by-four case, such a matrix might appear as follows:

Product Categories	*Process Categories*			
	Critical	*Major*	*Minor*	*Incidental*
Critical	(1)	(1)	(2)	(3)
Major	(1)	(2)	(2)	(3)
Minor	(2)	(2)	(3)	(3)
Incidental	(2)	(3)	(3)	(3)

The actions resulting from an actual situation corresponding with the above cells might be stated in this manner:

(1) Product and/or process redesign is required to reach, at worst, a cell carrying a (2) status; the existing or new process must also be brought into statistical control with at least 100 percent inspection of items produced for all status (1) characteristics until status (2) is reached in all such cases
(2) Either product and/or process redesign must produce a (3) status *or* a process control plan must be established that will ensure process centering within the tolerances and with no increase in variability—coupled with a program of continuing reduction of process variation until a (3) status is reached
(3) No action required, unless we are working in a continuous quality improvement mode when reduction of process variation is required irrespective of tolerance limits

Classification of Characteristics (C of C) can be initially confusing for design engineers with a strong background in functional design. Typical comments are: "Proper dimensioning and tolerancing communicate the relative importance of individual specifications" and "Proper dimensioning and tolerancing ensure that no specification has an Incidental design classification." To explain clearly why the technique provides information useful to enhance functional design, the classical decision theory concepts of "risk" and "loss" are often helpful.

Even though each feature has been toleranced based on product function, the relative consequences (loss) of equivalent nonconformances will not be equal. The design classification decision is the determination of the relative loss associated with a specified nonconformance. The Incidental classification is used to designate the lowest relative loss. As a result, it would be inappropriate to assume that nonconformances associated with specifications having an Incidental design classification do not matter. If specifications are functional, some effect will result from any nonconformance. The correct interpretation would be that characteristics with an Incidental design classification have relatively low cost associated with nonconformances.

Independent of the design classification, the likelihood of a process producing nonconformances (risk) is not equal for all specifications. The process classification decision is the determination of the relative risk of nonconformances associated with each specification. In the Critical case, the risk is unknown but assumed to be very large. The Major process classification category is used to designate the highest known risk of nonconformance. The determination of process capabilities provides feedback to engineers regarding the capability of common materials and processes. In time, this feedback results in designs that better support manufacturability.

The combination of risk and loss associated with the process and design C of C is used to help make decisions regarding required design specifications, process selection, process control, and associated documentation. Ideally, no worst-case combinations will exist where high risk is combined with high loss. Other undesirable combinations (high risk/medium loss or medium risk/high loss) may not be avoidable. C of C provides a method for accessing and communicating the combination of risk and loss associated with specific nonconformances. This "risk and loss" approach is the most effective method for prioritizing improvement action.

A key consideration of the application of a Classification of Characteristics program is that everyone must understand that a "minor" or "incidental" classification does not provide an automatic increase in tolerance. A supplier or an internal manufacturing unit must still meet all drawing and specification requirements; the former is specifically subject to the first lot and semiannual complete evaluation of 3.10 Supplier Rating Plans, and the latter is subject to the process control activities of 5.3 Received and Produced Quality Data Reporting.

Classification of Characteristics adds substantial additional guidance information to that normally provided by the designer, thus promoting much more effective resource application and understanding of the design intent by everyone—including, interestingly enough, the designer.

6.2 Classification of Nonconformities

When people talk about Classification of Characteristics, they often mean the relative importance of the degree of nonconformance of the characteristic rather than the relative importance of the characteristic itself. When they do, the discussion is really about Classification of Nonconformities. An example might involve the upper limit of a shaft length tolerance—say, 74.05 mm. That *characteristic* might be classified "high" because a small excursion, perhaps to 74.10 mm, would prevent assembly of the item with the mating part. But a lot might come in with parts measuring 74.06 mm, 74.08 mm, 74.10 mm, and 74.12 mm, among others. The .01 mm nonconformity might be classified "minor," the .03 mm nonconformity might be classified "major," and the other two "serious," since they could be corrected by rework.

When measurements are taken, the out-of-tolerance results can be viewed as more or less important depending on how much the tolerance is exceeded. Consistent with the discussion in 6.1 Classification of Characteristics, an excursion of 10 percent of the tolerance spread may be viewed as being of little cause for concern, but 50 percent could be very serious. With attribute data, varying importance may be attached to the type of nonconformity.

The terms used in Classification of Nonconformities ("very serious," "serious," "major," and "minor") are distinct from those used for Classification of Characteristics, but they have similar definitions. Those definitions might be as follows:

Very serious—That deviation which *will* produce or lead to a significant safety hazard to anyone, scrap, or a complete performance failure

Serious—That deviation which *will* produce or lead to some safety hazard; significant performance or reliability reduction; substantial field service or saleability problems; highly probable handling damage; or rework, repair, or downgrading of product

Major—That deviation which *may* produce or lead to minimal safety hazard; some performance or reliability reduction; some manufacturing difficulties; a limited effect on quality, serviceability, or saleability; or some handling damage

Minor—That deviation which *cannot* produce or lead to any problems of any nature

Using these definitions, the product designers, manufacturing, Quality, and service planners, and supplier personnel, where involved, can work together to develop limits for each category, preferably as a class, and the action to be taken in each case.

Classification of Nonconformities is undertaken to provide a consistent, economical basis for decision and action to be followed when a nonconformity is found. Also, by application of a logical demerit scoring program, operations can be rated and trends analyzed for improvement needs and accomplishments. A demerit rating scheme is illustrated in 3.10 Supplier Rating Plans.

6.3 Design and Analysis of Reliability and Safety Studies

In most product design and development organizations, it is customary to test one or more prototypes of a new or highly revised product to see if it fulfills its expected performance goals. In considerably fewer cases it is standard practice to test enough units under the full range of application conditions to determine whether the product satisfies all proper requirements for reliability and safety. The failure to determine the product capabilities in these areas stems from four sources: no preset goals for these characteristics, the large number of units associated with providing high degrees of statistical confidence of very low failure rates, a lack of sophistication in customers who do not require such assurance, and the substantial elapsed time for conclusive in-use tests, perhaps in the customer's hands.

External pressures are now forcing more and more manufacturers to re-explore what can be done to satisfy themselves in these areas. Product liability, insurance cost and availability, growing service and use cost-consciousness in the marketplace, and government action, particularly in the United States, all place a premium on demonstration of new product safety and reliability. Reliability prediction efforts, with associated goal settings, are being undertaken even outside the electronics industry, which for some time has had access to much data from military and other government experience for this purpose. Safety goals and controls are being established for many products to reduce exposure to product liability claims and to punitive action by government agencies.

The use of accelerated life or high-stress testing of relatively small numbers of units, combined with sophisticated curve-fitting techniques, can produce a highly reliable indication of failure modes and timing. In addition, computer simulation of product function under both normal and extreme conditions can, in a much shorter time than the field tests of the product would take, provide much useful guidance to Design for preventing safety hazards and other failures. Such tests must be regularly compared with actual field experience to validate acceleration factors and to show that the failures experienced in test are found also in the field. If such correlation is not demonstrated, changes in the test programs or interpretations are indicated.

Customers are becoming more sophisticated and demanding, through technological and consumeristic education and exposure, and are quicker to react when their demands are not satisfied. Producers fail to recognize and react to these forces at their peril.

Therefore, the need to assure oneself of the reliability and safety of the product, both before it is introduced to the market and from time to time during the product cycle, is increasingly more important. The setting of reliability and safety goals, the design of the product for testability for both reliability and safety (see 2.4 Reliability Planning and Incorporation and 2.5 Safety Engineering), economic test planning and execution, the use of computer simulation, the careful control of the validity of tests through knowledge of the items being tested (see 2.13 Preproduction Testing) and verification of test parameters, the thorough analysis of results, resultant design modification when indicated, and the validation of conclusions through comparison with experience are all implicit in the decision to obtain the necessary assurances that one's product is reliable and safe.

When, through Design and Analysis of Reliability and Safety Studies, these tests are properly designed and run and the results incorporated in the product design, management can be much more confident of customer satisfaction and of having reduced the likelihood of successful product-liability lawsuits.

6.4 Environmental Impacts

Recently, there has been a growing recognition of and concern for the effects of industrial practices and products on the environment; but as yet, little is understood about the subtle interrelationships between mankind and this planet—although much is suspected. Some of the gross effects of our activities have been identified and, when the cause has been removed or reduced, improvements in ecological health have resulted. Therefore, governments and citizen groups have moved to regulate or modify the impact of industry on the environment, often with the strong cooperation of the industries involved.

Some hysteria has accompanied this effort, resulting in an occasional cure that is worse than the disease or is economically unjustifiable when compared with a viable alternative. The Quality System is involved in several aspects of the entire process, including its contributions to applying sound experimental considerations in helping to determine valid cause-and-effect relationships, and thus it is fostering a more rational approach to these emotionally charged issues.

Another concern less commonly addressed today is the reverse of the public one: What effect does the environment have on the product and the processes used to produce it? Sometimes the pollution caused by industry X or product Y may prevent company Z from producing *its* product successfully or profitably. For example, some time ago certain fishery products were unsaleable because of high retained mercury levels, reputedly from industrial waste; whatever the truth was, the potential relationship existed, and the control of industrial waste disposal received an additional incentive.

There are many similar examples—photographic film damaged during manufacture by airborne particulates such as fly ash and radioactive particles, fabric colors affected by unwanted trace chemicals in the dye bath, cleaning processes rendered ineffective by dust, and others. Those responsible for the Quality System must be alert to potential hazards of this nature and take steps to prevent their having significant effect.

In addition to attempting to avoid process contamination by studying the relationships between the process steps and the environment, it is also necessary to determine what the conditions of distribution, handling, or use will be for the product. Exposure of photographic film to x-rays or other radiation may make it unusable unless suitable shielding is provided in the package. Vibration or shock, weather, magnetism, voltage, temperature, wind, humidity, and chemicals are just some of the factors that can affect the ability of the product to perform safely and satisfactorily.

It is important also to consider the limits of variation in these factors within which the product will be used and the excursions from them that would constitute "reasonable abuse"—use of a product for purposes other than intended under conditions not normally expected. If the designer fails to provide for such abuse, he risks customer dissatisfaction and, possibly, product liability claims. The designer must also consider the duration of the probable exposure to severe stress. For example, a voltage "spike" may have an entirely different effect on the product's performance or durability than prolonged exposure to even slightly higher than normal voltages.

Thus, to perform satisfactorily in this complex area, the company must ensure that its products and processes are protected from the undesirable effects of the environment, including pollutants from other companies and nature, and at the same time are prevented from damaging the environment. Deleterious effects on the environment include the effects of disposal or escape of chemicals, thermal and other radiation, noise, disturbing the earth, and others.

Given either regulatory or internal requirements in this area, a company can capitalize on its positive efforts to comply in many ways. Frequently it is found that controlled, clean factories, processes, and products produce an overall cost reduction. Employee turnover, absenteeism, and grievances are often reduced. Acceptance by the community is enhanced, with the advantages accruing which that implies, if the work and its results are discreetly publicized. Positive, effective action in critical areas can often forestall more expensive regulation encompassing unimportant areas as well.

6.5 Machine and Process Capability

An extensive literature exists on how to conduct machine and process capability studies. Such studies are fundamental to quality control because it is implicit in any statistical quality-control program that we learn how our production equipment and processes can perform compared with the product tolerances they are supposed to meet. These studies give the producer a basis for obtaining tolerance changes, perhaps as trade-offs, to reduce or eliminate the production of nonconforming items. Alternatively, they can provide a clue on where and how to improve the process to achieve a similar result. In addition, they are the basis for judging the effects of reducing variability of processes to zero. The literature provides the mechanics for performing these useful studies under virtually all circumstances, both directly and indirectly, so that necessary corrective action can be taken.

A preventive, rather than reactive, attribute of capability analysis, however, is that such information can be used as a control in the design of a new product. Guided by the manufacturing and supplier part of the Classification of Characteristics (6.1) program, designers can establish product yield goals that are realistic with respect to known processes. Such knowledge promotes design decisions oriented to cost optimization, when coupled with information related to the other cost ingredients. It also identifies the areas of opportunity for continuous quality-improvement efforts.

Machine capability is a measure of the inherent ability of a piece of production, measurement, or control equipment to reproduce its results with accuracy and precision. Process capability similarly measures that machine in its operating circumstances—with variations in the material used, the personnel operating it, the instruments associated with it, the ambient conditions surrounding it, and the other factors that, with the machine, form the process.

In developing a capability study and use program, we must weigh every aspect of its application as it contributes to overall quality capabilities. This means we must specify the degree of capability of new equipment and processes and demonstrate that capability, so that the equipment supplier can produce it and we can test it at his factory as well as in ours after installation. We must assure that the suppliers' and our own key manufacturing processes can produce a new item, both in the product design yields and in the demonstration of continuing and improving capability as production proceeds. In every case, we are seeking an understanding of the product and process interaction that will permit us to optimize our overall production and test costs. This program does not stop when a new product is in production: New materials or production processes introduce new factors into the optimization determination; new or modified fixturing and material handling equipment have similar effects.

Direct capability studies are those performed in the manner of, and frequently using the techniques of, statistically designed experiments. Indirect studies occur when, for product acceptance or process control purposes, we analyze routinely obtained inspection, test, and maintenance records or those obtained from preventive or breakdown maintenance activities. Both approaches should be used as appropriate. The use of indirect studies often leads to a meritorious direct study. All studies should include a comparison of the results with a preestablished standard, the violation of which produces a change in the product design or in the process.

The use of a Machine and Process Capability study program in its entirety, coupled with the work of 6.7 Quality Level Improvement, results in an affirmative answer to the question "Can the manufacturing processes produce the item with negligible losses or repairs and, ultimately, with zero variation?" This goes a long way toward optimizing total production costs.

6.6 Problem Solving

In all parts of an enterprise, from the office of the head of the organization through all levels and functional areas of management down to the newest employee at the starting grade, there is a need to identify and solve problems. Within the Quality System we provide for this to be done in 1.12 Corrective Action Program, in 6.7 Quality Level Improvement, and in 6.8 Statistical Technique Application.

Whether the problem to be solved is one associated with the product or service provided to the internal or external customer, with the process involved in providing that product or service, or with some broad business decision such as where to locate a new facility, a variety of techniques may be usefully applied to the solution process.

Some of these techniques call for involving groups or teams in the solution activity. Others can be applied by an individual, even with or without advice from others. Devices include varieties of Brainstorming, Cause-and-Effect Analysis, Paretoization, Experimental Design, Value Engineering, Work Flow Analysis, and the use of other statistical tools.

In any case, almost always one or more techniques are available that have the ability to reach a solution better in every respect than that of acting upon the first idea to surface. These techniques are able to achieve that characteristic of the solution best described as "elegant"—as defined in 2.9 Product Design Review.

When the organization applies structured problem-solving methods to its decision-making processes, its personnel solve more problems more rapidly and with greater effectiveness than using any other approach. As a result, the conditions of the Total Quality Management effort and the Malcolm Baldrige National Quality Award guidelines are both addressed in the best possible manner.

6.7 Quality Level Improvement

There are essentially three reasons why a company might decide to improve the quality of its products or processes: because present levels are unsatisfactory, because of a commitment to continuous quality improvement, and in preparation for more demanding requirements. The third possibility arises from a product standpoint if, for example, the company is moving from the production of carpentry tools with few defined nonconformities to military equipment with many tight tolerances, or to a regulated product from a nonregulated one. From a process standpoint, the company might be introducing, to meet competition, a super-finishing process that demands tighter controls over rough and finished turning operations than the previous process required.

The indication that present quality levels are unsatisfactory may come from observed reject rates, field failures, regulatory or competitive changes, customer complaints, quality cost analyses, various audit results, or changes in management philosophy. In the context of continuous quality improvement, every process quality level is unsatisfactory unless the variability is zero and the value generated is at the design nominal.

Whatever the source of the decision to improve quality, the steps to be taken are the same for process improvements and, in many cases, the improvement of outgoing product. One step involves product redesign or Tolerance Partitioning (2.7) to eliminate sources of nonconformity. Another involves process replacement with a more capable process or part of the process, or by adding portions to the process. This could also include mechanization or automation of the process to minimize operator-caused nonconformities and reduce variation. These activities are normally initiated by management after Machine and Process Capability (6.5) studies have identified the product design or the processing equipment as the cause of the inability to satisfy present or expected quality requirements or to decrease variability.

A major source of quality problems involves the operators. In a very few such cases, the operator is solely to blame. In most, management contributes to or causes the problem. The list of management failings in this area is staggering; it includes, among others:

Improper selection and training of the operators
Ineffective transfer of specific work instructions and standards information to operators
Inefficient work place layout and materials quantity and quality provisioning
Supervisory refusal to stop production to correct quality problems before they become disasters because they feel that meeting production goals is more important than quality
Ineffective production tooling and equipment
Insufficient and incapable maintenance of equipment
Failure to provide self-inspection capability and feedback of quality information to the operator
Insufficient technical support when a quality problem arises

To sum up this list, the reason the great majority (some say at least 80 percent) of operator-involved nonconformities occur is that management has failed to support the operator or the process adequately. Since many of these nonconformities may be identified to specific parts or locations on the product, their frequency over *all* parts or locations may go unnoticed by management, and no corrective action is initiated. In fact, in many organizations, action may be taken only when problems become catastrophic. This situation of many different nonconformities with few repeats of each has provided the area where operator participation in problem solving has been most effective.

By involving the operators in the process of defining and resolving the problem, companies can supplement their limited technical resources with a large body of interested contributors to quality improvement. With training in problem-solving techniques, applications of industrial statistics, and group dynamics, groups of operators can become virtually independent of technical support in determining the basic causes for quality problems and the best approach for eliminating them. The Zero Defects programs of the 1960s, the Japanese Quality Circle concepts, and other successful approaches to participative quality improvement have all depended on providing these four basics: the methods for achieving improvement, the opportunity for operator involvement, a mechanism for showing that the improvements result from the specific improvement project, and the management commitment to eliminate the causes of quality problems. (Further development of this subject is provided in 8.10 Personnel Quality Participation.)

When these, particularly the last, have been provided, results have been dramatic. An interesting side effect of these programs has been that when solutions have been developed astutely, many problems, instead of requiring separate solutions, have disappeared with one corrective action. This shows that our methods of identifying nonconformities frequently fail to recognize their interdependence, and thus their real magnitude. This interdependence is commonly discovered when correcting the cause of an apparently "internal-only" nonconformity results in the disappearance of a field failure whose source was unknown. For this reason, many highly successful programs start with eliminating nonconformities associated with purchased material and correcting the sources of nonconformities while moving step by step through the manufacturing process. It is common to discover that all rejects of final product have disappeared by the time one reaches that stage of the investigation.

The approach within a department might be to list the problems in order of frequency, correlation with field problems, or total cost; assign the top five for attack, and add to the active list as the worst are disposed of. When the problems prove to be wholly operator-controllable, provide specifics, including samples or posters, showing the right and wrong techniques and results. When successful results are obtained, the operators involved should receive public recognition for their contributions.

Whatever methods are adopted for Quality Level Improvement, whether one is trying to improve yields of classes of equipment, to reduce production of nonconforming material by suppliers or operators, to improve quality that already meets specifications, or to upgrade process capabilities for handling a more demanding product line, the concepts are the same: Define the problem, develop a plan for dealing with it, find its cause, eliminate it. While doing this, use the knowledge and talents of all involved personnel, particularly the operators, to help achieve the desired result. This multiplies the available problem-solving resources, reduces the elapsed time for solution, and, probably most important, has a significant positive effect on operator morale and quality motivation.

As is true in any corrective action activity, it is important for anyone doing Quality Level Improvement work to make sure that the changes introduced do not cause problems elsewhere. Assurance of the lack of undesirable "downstream" impact may require controlled experimentation before the change can be instituted. Such assurance must be supplemented with later observations or measurements to determine the net effect of the improvement, coupled with any necessary process controls to prevent unplanned excursions from the new process.

The benefits from this program in conjunction with 1.12 Corrective Action Program, 4.13 Yield Control Program, 6.8 Statistical Technique Application, and 8.10 Personnel Quality Participation, and the others referred to above should be obvious: improved quality, reduced costs, improved work force morale, and increased quality capability and capacity.

6.8 Statistical Technique Application

The theory, description, and use of statistical quality control techniques in business, industry, and government have been thoroughly covered in many texts. It is not the purpose of this book to repeat those detailed descriptions. This element, Statistical Technique Application, is oriented toward making sure that these powerful devices are fully understood and appropriately used by the enterprise. All members of every organization, from the chief operating officer to the worker on the factory floor (including those in non-product-related functions), should have a sufficient knowledge of the pertinent techniques, should be able to interpret the results of a statistical analysis or presentation correctly, and should know whom within the organization to contact to obtain expert assistance in developing, interpreting, and applying the techniques.

Over the more than half a century since modern industrial application of statistical techniques began in the Bell System, numerous organizations have used them, usually through individual initiative, and have thereby profited. Many times, when the individual who made the often startlingly successful application of control charts, sampling plans, designed experimentation, regression analysis, or some other device moved on, the techniques either disappeared from the enterprise or were improperly and ineffectively used. Some companies that achieved international recognition for leadership in the application of statistical quality control techniques, particularly in their explosive growth period during and following World War II, had abandoned them completely only a few years later, despite the very positive benefits they had previously enjoyed. Investigation of a number of these reversions showed that the collapse followed the death, the departure, or occasionally the transfer or promotion of the program's initiator.

The great advantage of the Quality System approach is that, once the successful use of these techniques is begun, this element and some of those in Subsystem 8 require a continuous education effort (supported by Quality System controls such as Audit) of such magnitude that it becomes virtually impossible for the programs and associated benefits to be lost.

Another aspect of this element is that the programs should be examined regularly for continuing utility. Sampling plans may be replaced by others or by automatic inspection of 100 percent of incoming or produced items at reduced cost and with increased production. Control charts might usefully be changed from $\overline{X}$ and R to Median and Range, or from p to pn, or be generated by computer. In any event, they should be removed and replaced periodically once solid control is established at a satisfactory level. Designed experiments may be supplemented by participative problem-solving techniques with accompanying employee morale benefits—in the style of Quality Circles, Kepner-Tregoe, and others. Such regular revitalization of the programs is necessary for their continued acceptance and effective use.

Proper Statistical Technique Application ensures that the organization will obtain and continue to realize the benefits of these often misunderstood, rarely properly appreciated, but very valuable tools.

6.9 Unique Delivery Requirement Management

Most manufacturing organizations and some service groups are frequently faced with nonroutine product or delivery requests by customers. These often pose no special risks for quality, reliability, or safety of the product—but they sometimes do. The Quality System approach requires that these transactions be viewed as potentially of high risk and that appropriate steps be taken to minimize such risk.

While the Unique Delivery Requirement Management element assumes that the customer's special requirements will be satisfied whenever practicable, it also recognizes that nonstandard practices may be necessary to achieve this result safely. Sometimes the unique product characteristics demanded may induce a new failure mode or rate for a nearby standard component; the special feature may not, of itself, be as reliable as the basic product; or the change might result in the product's being unsafe under certain usage conditions, with resulting increased product liability. These possibilities dictate a Product Design Review (2.9) on an accelerated schedule, perhaps of a type different from normal.

The unusual or changed delivery requirement might be for extended delivery, which can introduce handling, storage, and shelf-life questions, but more likely it will be for shorter-than-normal lead times. For these, material substitutions may be required, time-dependent processes may prove critical, time-consuming testing may have to be simulated or may subject the shipment to field repair on a lot sample basis, or field completion of the product may be necessary. All of these conditions imply a special appraisal effort and other possible control applications.

Since these extraordinary efforts may well add cost to the operation, they should be planned for and the cost recovered from the customer, if practicable. Regardless of who pays for such activity, however, the requirement for meeting the customer's needs must be satisfied while simultaneously achieving the objectives of the Quality System. This may require cooperation, perhaps even concession, from the customer, as well as support from all involved company components.

Protecting the company and the customer under these circumstances demands considerable flexibility and, frequently, imagination from the organization. These capabilities must be exercised within the control of the Quality System to avoid potentially serious consequences to cost and quality reputation.

Subsystem 7

Quality Measurement and Control Equipment

The intent of the Quality Measurement and Control Equipment subsystem is to provide an economical yet effective program to guarantee the continuing effectiveness of the measuring and control equipment that is a vital component of any Quality System. The equipment covered by this subsystem includes everything used to measure or directly or indirectly evaluate parts, components, subassemblies, assemblies, and products, and the tools, jigs, fixtures, processes, and measurement and control equipment used to produce or to measure them—regardless of whether the equipment is located in a laboratory, factory, warehouse, office, shop, cabinet, or desk drawer, and regardless of to whom it may belong. It also includes automatic measurement, feedback, and adjustment mechanisms and other indicating devices associated with production equipment that may or may not involve computers. All of these categories might be represented by manually operated or automatic equipment that may be active or passive in operation. Subsystem 7 covers all of them when they are used for product or service acceptance, product or process development, process continuation or adjustment, or other business decisions.

Obviously, the programs associated with such a large volume of equipment must be carefully tailored to the real needs of the organization, including that for long-term reliability of the equipment, and with close attention to the economics of the situation.

The elements are arranged in the order in which they might be applied in a new operation.

7.1 Quality Measurement and Control Equipment Plan

In the view of many quality professionals, one of the most neglected areas of efficient application of resources is that of equipment for product (including for preproduction or development engineering activities), process, field installation, and service control and appraisal. Whereas we may want to advance the state of the art in our product technology and, as a result, are sometimes forced into a similar posture for manufacturing, we frequently refuse to consider changing our ways in inspection and test. Consequently, we often do not invest as we should in advanced control devices for our manufacturing equipment, thus perpetuating a dependence on outmoded appraisal equipment and practices.

Since the quality-related equipment under discussion may cost a lot of money and take a long time to develop, obtain, and install, it is necessary to have an orderly plan for its acquisition. Done at a sufficiently early stage, this provides enough time to obviate the all-too-frequent condition of late recognition of need, resulting in the decision to muddle through with existing capability. An organization that deals properly with this subject will maintain a continuing knowledge of new developments in the field, and sometimes participate in providing such new developments. Not every company needs a development laboratory of this type; the ones who can benefit from such effort should invest in one and others should develop close contacts with the commercial houses that produce such equipment for sale.

When an impending need is identified, either type of approach will enable the company to make the proper decisions with respect to measurement and control concepts, capabilities, capacities, costs, compatibilities, and process and product characteristics to be involved.

The concepts to be considered include manual, mechanized, or automatic gauging or testing as well as passive indication versus automatic computer-aided feedback controls, direct versus indirect measurement, inline or off-line applications, and destructive versus nondestructive evaluation, among others.

Capabilities include appraisal, control, and response times, durability, measurement sensitivities, measurement accuracy, operating ranges, any calculations or analyses required or performed, current or pending regulatory requirements, competitive considerations, and the number of different characteristics subject to control.

Capacities are related primarily to agreement with planned production rates, sampling versus 100-percent evaluation, and ability for expansion.

The costs to be covered are development, acquisition, operating, and those for necessary alteration and are judged in return on investment, later savings, and trade-off contexts against other costs and technical considerations.

The subject of compatibilities opens some new dimensions to the study: environmental questions; the existence of sufficient resources to operate, calibrate, and maintain to established schedules; the availability of adequate utilities, supportive facilities, and capable technical support; and the match with companion equipment (supplier, internal, or in the field, as described in 3.3 Supplier Measurement Compatibility and 9.4 Field and Factory Standards Coordination) so as to solve rather than to induce problems.

The determination of which characteristics are to be included involves selection according to the categories described in 6.1 Classification of Characteristics for control by process or by appraisal, astute combination of characteristics in a single control device or method, and provision for alternate methods of control in case of breakdown or ineffectiveness of primary approaches.

Once all of these questions are satisfactorily answered, we can proceed to develop or to procure the needed equipment. In those cases where the equipment is uniquely developed or is sophisticated, its capability should be determined as described in 6.5 Machine and Process Capability. This should be done when practicable at the supplier's location, but, in any event, as the result of a proper definition of requirements and demonstration methods in the procurement documentation. Part of the equipment installation process includes the timely furnishing of all useful operating, safety, calibration, and maintenance manuals and other necessary documents in a consistent form and format most understandable by the users. These will be controlled as specified in 7.4 Quality Measurement and Control Equipment Documentation.

When the Quality Measurement and Control Equipment Plan is properly prepared and followed, measurement errors are reduced, time is saved, quality and other costs are reduced, and product and service quality, both internal and as seen by the customer, are improved. Other recognized benefits from the program include reduction in unprofitable discussions over "nonconformance," in quantities of nonconforming items produced, and in carrying charges for "debatable" inventory. This activity may also contribute to 1.10 Quality Inputs to Advertising.

7.2 Quality Measurement and Control Equipment Safety

The equipment used to inspect and test product and to control manufacturing machines and processes is often looked on as much less dangerous than the manufacturing equipment itself. To a large extent this view is valid, but it contributes to an attitude among operators, calibrators, and maintenance personnel that can be very dangerous to them and to others.

Some of this test and control equipment operates with high internal voltages; other items have significant potential for crushing or cutting the unwary. To minimize the safety hazard, the equipment should be designed for negligible risk of injury; electrical interlocks and positive mechanical stops should be made integral with the equipment; safety instructions, regulations, bulletins, illustrations, and cautionary placards should be provided; and personnel working with the equipment should be carefully selected and trained to exercise appropriate care in their activities.

The correct amount of this type of information should be provided, in a format aimed at attracting the attention of the operating and servicing people. This implies the orientation of the material to the educational levels, cultural considerations, and native languages represented by the involved workers. Testing of the material before its release is necessary, and audits of its effectiveness in use provide the basis for its continuing improvement in utility.

7.3 Quality Measurement and Control Equipment Facilities

Most of the other elements in this subsystem assume the existence of equipment for quality measurement and control. This element is intended to ensure that such equipment is appropriately housed, protected, stored, handled, and used to obtain long life and valid results. Quality measurement and control equipment is used, for example, in laboratories, in factories, at installation sites, and in repair facilities—with environments varying from ocean bottoms to outer space. The same or similar equipment is provided in several grades, or accuracies, dependent on the type of use for which it is destined.

Each higher grade imposes greater demands upon its users for more stringent and stable conditions for its use—conditions of temperature, humidity, vibration, cleanliness, calibration capability and sophistication, and so on. Many organizations have calibration laboratories of their own, traceable to national or international standards agencies, where they calibrate, maintain, and store their working gauges and other equipment. (See 7.6 Equipment Calibration and Maintenance.)

Even the working equipment often requires temperature, humidity, vibration, or cleanliness conditions that are better than the normal factory environment to which it may be exposed. Parts of or entire factories have been air conditioned, vibration isolated, or made into "clean rooms" solely because of the relative delicacy of vital equipment of this nature. Interestingly enough, when this has been done in older factories, it often has produced unexpected dividends in improved quality, employee turnover, and overall costs (see 4.12 Housekeeping) outweighing the costs of the change in working conditions.

The point of this element is that it is stupid to invest in expensive equipment to perform required inspection, test, or control functions and then fail to support it physically so that it can accomplish its intended purpose. This is true particularly when management expects to make important decisions, perhaps affecting their customers, based on the results obtained from such equipment. (See 1.8 Product Liability Prevention.)

7.4 Quality Measurement and Control Equipment Documentation

As part of the ability to calibrate and maintain properly the inspection, test, and manufacturing process control equipment, it is essential that all pertinent documents be available and current. This means that they must be carefully filed, updated when changes occur, returned to the file expeditiously, and replaced as needed. For an organization with a lot of such documented equipment, this can be a considerable chore. The costs of trying to cope without providing effective clerical effort in this area are, however, much greater—although they are concealed in loss of maintenance efficiency rather than visible in clerical costs.

The list of involved documents is extensive. It includes operating manuals, service manuals, drawings, specifications, schematics, bulletins, parts lists, standards, and instructions, among others. Appropriate indexing to equipment number and location makes the information more readily accessible and therefore more useful. Much of this mass of documentation is provided by the manufacturer of the equipment. Some, including that required for home-built equipment, must be internally generated.

In any case, a formal Quality Measurement and Control Equipment Documentation program should ensure that the necessary documents are obtained and entered into the control activity. When this is successfully accomplished, quality problems and associated costs arising from improperly or inefficiently calibrated and maintained equipment will be significantly reduced.

7.5 Quality Measurement and Control Equipment Applications

As new and modified products are being prepared for production, and when new manufacturing equipment and processes are being purchased or developed or old ones are being changed, it is important to examine the equipment currently in use. The equipment involved is that used for laboratory, incoming, process, and final inspection and test and for control of equipment. The examination should explore equipment capacity, capability, and compatibility with these new or changed items. Sometimes we will find it economical and efficient to use our current equipment, perhaps with minor modification or with some increase in the amount of such equipment. At other times we may have to make major changes to the present equipment—adding automatic material handling capability to meet increased capacity demands, for example.

Wherever it is practicable, we should lean toward the use of existing equipment to simplify the logistics of calibration, maintenance, and parts supply and to minimize the training and retraining needs of the organization. In following this principle, however, we should not use it as an excuse for failing to obtain the most efficient equipment. Appraisal costs are frequently excessive just because people do not search for the best measurement and control equipment to use, so we always should be alert to opportunities for improved operations in this critical area.

During Product Design Review (2.9) and Process Design Review (4.1), those having the measurement and control equipment planning responsibility can become aware of forthcoming products, processes, and machines that could require the work of this element. Definite programs should be instituted to ensure that advantage is taken of such opportunities to identify specific matchups of measurement and control equipment with product characteristics and process equipment.

When, in the course of the product development cycle, we have produced major experimental quantities—as frequently occurs in preproduction and pilot runs—we should evaluate from several viewpoints the measurement and control equipment used. One question to be answered is whether it accomplished its objectives: Did it inspect or test correct characteristics, giving accurate readings at the necessary rate and with sufficient economy, or did it stabilize the processes so as to yield only satisfactory quality? Another question relates to its future application in development: Should we plan to use it for other new product programs? The third question is not always pertinent, depending on the extent to which the trial run reflects production: Should we use the same equipment, or a production version of it, in full-scale manufacturing? If the answer to any of these questions is negative, we should initiate corrective action to ensure that the organization benefits from this evaluation.

The same concept is applicable to equipment usage in production. We should regularly, typically annually or semiannually, evaluate the Quality Measurement and Control Equipment Applications and the proper use of the equipment. Taking steps to remedy any indicated shortcomings will promote operating efficiency, effectiveness, and economy.

7.6 Equipment Calibration and Maintenance

When new inspection, test, and control equipment is installed, it is often satisfactorily calibrated by the manufacturer or installer. But, even if it was satisfactory to begin with, the equipment tends to give increasingly less accurate results with use and time. Users of such equipment recognize the need for regular examination of its capability and frequent adjustment to produce correct answers and signals.

The calibration examination measures the equipment's controlled conditions. When the response is sufficiently incorrect, the equipment must be adjusted, repaired, or removed from use. This type of maintenance is part of the overall maintenance program, which also includes preventive and breakdown maintenance.

It is immaterial to the program whether the required calibration and maintenance work is performed by company personnel, by outsiders working in company facilities, or by outsiders at an external location. The relative overall costs would normally provide the basis for choice among the alternatives. The maintenance and calibration effort also requires substitute equipment, a working schedule based on use effects for greatest economy, and a positive action program that ensures the schedule is met.

An in-house program requires special equipment, standards, proper facilities, and sufficient trained personnel for both calibration and maintenance. Any program, whether or not in-house, requires properly stored and up-to-date schematics, instruction manuals, and other documents, and complete records of relevant transactions; these requirements are discussed in other elements of this subsystem. A calibration program for the normally more sophisticated special calibration equipment and standards is another need.

New, modified, or repaired equipment should not be released for use unless all necessary calibration has been performed, nor should anyone use equipment that does not carry the proper notice of current and satisfactory calibration.

This program, which in many companies is limited to production and related inspection and test equipment, must also include the equipment in research and development laboratories, that used by suppliers in producing material for the company, and that used by personnel in the field during installation and service. The form the program takes in dealing with outside organizations may have to be different from that for the internal one, but the company should not overlook the possible benefits of assurance and coordination gained from providing such a program to those outsiders.

In addition to the planned schedule of calibration and maintenance, the program should include study of and reaction to process control results or process capability analyses that may indicate involved equipment problems. This validation aspect of the program would include supplier data, receiving inspection results, field complaint analyses, audit findings, and so on. Such evaluation may result, for example, in changing calibration schedules, instituting operator checks on equipment, or developing different equipment requirements for replacement items.

A question often arises as to the percentage of out-of-adjustment items found on calibration that would mandate a change in calibration cycle. An evaluation of the economic impact of the implied erroneous product or process decisions would be the best basis for this judgment. Because eventually any unadjusted item will fail a calibration check, it is also necessary to include the past history data in this percentage: How many preceding cycles did the adjusted items go through before this failure? Certainly, if a gauge characteristic is always or frequently in need of adjustment, the calibration cycle time must be reduced. As a rough guide, there is some opinion that if, within one cycle, as much as 5 percent of a given type of equipment require adjustment when calibrated, the cycle is too long. Each organization should set its own criteria for cycle adjustment, and the criteria may be different for different types of equipment.

A proper Equipment Calibration and Maintenance program will increase the probability that correct quality and manufacturing decisions are made, with resultant savings in cost and time.

7.7 Audit of Quality Measurement and Control Equipment and Usage

Organizations that use this type of equipment to assess the performance or the conformance quality of their products during design and development of their purchased material, of their material in process, and of their finished product and use it to control their manufacturing equipment usually have it calibrated to ensure that it continues to give "right" answers. Often they become complacent with the protection afforded by the calibration program and overlook the possibility that with time, if not at the beginning, the equipment may be accurately measuring or controlling the wrong thing.

The audit specified here is oriented to uncovering both types of failures—those arising from inadequacies in usage and in the existing calibration and maintenance program and those reflecting an incorrect application of equipment. In the latter case, the audit can frequently detect impending misapplications if the auditor maintains an awareness of upcoming new products and processes, thus serving to help prevent the problem from arising. Modernization or replacement of equipment, then, is a logical outgrowth of this type of audit.

The audit plan will be somewhat different from that described in 1.14 Audit of Procedures, Processes, and Product in that it will be much more intensively oriented toward the examination of the proper application of the equipment, rather than toward measuring the degree of strict conformance with stated requirements. Obviously, it would require audit participation of personnel who can properly make the decision whether the correct characteristics are being measured or controlled. This will probably require assignment to the audit team of specific individuals from certain groups rather than permitting randomly selected representatives from such groups to participate. Otherwise, the audit should be conducted in a manner similar to that covered in the System Audit Element (1.14).

Other differences may also arise in reporting the audit results, since trends in "incorrect application" experience may not have any pertinent meaning. This will have to be decided on a case-by-case basis. Of course, trends in deficiencies of the calibration and maintenance program and in usage nonconformances always would be useful control measures.

The Audit of Quality Measurement and Control Equipment and Usage provides continuing measurement of the validity of equipment selection and application and continuing indication of any needed training or direction of personnel operating it.

Subsystem 8

Human Resource Involvement

Human Resource Involvement addresses the need to obtain full commitment from all personnel to doing all the things necessary to achieve the Quality Objectives. Without such participation, the Quality System cannot operate. In addition, this subsystem is concerned with satisfying the need for growth of all involved personnel.

The Quality System can be successful in producing customer satisfaction at a proper profit only if it is operated by knowledgeable, capable people who want to make it work. It involves a fundamental premise: that people want to do a good job. In my experience, the only time that is not true is when there has been a previous history of poor management applied to the individual; sometimes the true source of the problem is the president of the company.

An illustration of the effects of bad management might be of interest. A high-school graduate took a factory job operating a machine. After a time the machine began to run poorly, producing increasing quantities of nonconformities. On several occasions she brought this matter to her supervisor's attention and was told to "get back to work, we need the parts!" Finally the machine broke down completely and was put back into active but nonconformity-producing service by an inept maintenance man. The employee's subsequent protests were firmly silenced by the supervisor. She swears that for the remainder of her industrial career she will never again concern herself with quality. (See 8.10 Personnel Quality Participation.)

To overcome the effects of such destructive influences is difficult; the intent of this subsystem is to attack the problem from several directions—

proper training for all, positive development of personnel, and continuing provision of useful information about Quality System matters. For training to be useful, we must determine the specific system-related education and training needs of all personnel, from the top down. Then we can identify the content of the course(s) needed by each group and provide them on a planned, continuing basis.

The elements of the subsystem are organized into these three groups: eight for training, four for development of personnel, and two for information dissemination. The training and information groups are internally organized by recommended order of application; the development group is in alphabetical order.

8.1 Overall Training in Quality System Subjects

In some manufacturing companies, the top management personnel have significant experience in the successful application of statistical quality control techniques. Where this is true, they have generally seen to it that all of their subordinates receive some training in the subject. In many Japanese companies this idea is rigorously applied, with many board chairmen and presidents receiving forty or more hours of formal training in these subjects and shop operators receiving as much as 120 hours of formal quality-related training. Therefore, virtually every employee in the company can do effective quality engineering work when it is useful or necessary. Sales representatives, service representatives, buyers, employee relations personnel, draftsmen, and maintenance personnel are able to apply quality control techniques and Quality System concepts to their own situations just as well as can the design engineers, manufacturing engineers, reliability engineers, quality engineers, operators, inspectors and testers, and all supervisors and managers.

The potential benefits of such a high degree of quality proficiency in an organization are so enormous that the Quality System reflects this desirable situation quite strongly. Specific training programs are covered in several elements, such as 3.7 Quality Engineering and Training for Suppliers, 8.2 Management and Technical Personnel Training, 8.3 New Operator Quality Familiarization, 8.4 Inspector and Tester Training, and 8.5 Field Personnel Training. This element covers the training of a company's general population—those people who require an understanding of the objectives of the Quality System and of its programs, procedures, and techniques that they can use to improve the quality of their efforts and to evaluate scientifically the risks of the outcome from their programs.

Since the Quality System is the result of mutually supportive activities of many different functions, the training, in addition to seeking the objectives mentioned above, must promote team consciousness and multifunctional participation in quality programs and problem solutions. To increase and then sustain its effect, the training may be given in discrete, directly usable segments with required applications interspersed among the introduction of new items.

To be most effective in providing the proper type and amount of training to the correct people, the organization must conduct a training needs assessment—involving input from all useful sources (shop personnel, technical staff, supervision, suppliers, service providers, etc.) in addition to the groups directly involved in providing the training. This assessment, conducted annually, will determine resources needed, subjects to be covered, and priorities for all individuals and groups within the company to receive the necessary education and training.

Subjects covered include the concepts and procedures of the Quality System, management of the System through its control mechanisms, and selected statistical control techniques, as well as the material related to the other subjects included in the term "Quality" as mentioned in the Preface. Specific attention should be given to covering all appropriate material related to health and safety—such as safe work practices, handling of hazardous materials, human factors in design of products and processes, and environmental considerations.

When the majority of personnel in the organization, through Overall Training in Quality System Subjects, can apply Quality System concepts properly, the results will be evident in improved costs, effective reaction to internal and customer problems, and better quality levels throughout.

8.2 Management and Technical Personnel Training

Many of the elements of the Quality System involve training or are dependent upon training for their proper performance. Some of them are oriented to training as a subject in itself; others are identified to a specific subject or to a specific group of people. This Management and Technical Personnel Training element is one of the last named and is oriented to producing that Quality System managerial and technical climate within which the other training programs can be effective.

We need to be sure that all managers, engineers, and other technical personnel (including those operators who are engaged in supportive activities such as Quality Circles), whether or not they are part of the formal Quality organization, are sufficiently apprised of the detailed workings of the Quality System so they can understand and perform their associated responsibilities. They also must be knowledgeable enough in statistical quality control methodology to use it when appropriate and to comprehend its use by others. In addition, since many of the Quality System programs call for multifunctional group efforts, the managers and technical personnel must know how to manage and to participate in such interdepartmental projects.

Since the training must be tailored to suit the needs of the individual as well as those of the organization, a method for comparing the latter with the state of knowledge of each person or group is needed. The knowledge gaps, then, must be closed by the training program. This is not an absolute, as people when they are part of a group of students will tolerate some overlap with what they already know; too much overlap, however, is oppressive, and the new material is lost as a result of boredom.

Quality System concepts that might be applied to other business areas are particularly worthwhile exploring in these training sessions. Such practices as systems engineering, statistical analysis of data, the use of audit for control purposes, controlled sampling to examine complex situations and those when it's difficult to obtain data, techniques for problem solving, corrective action considerations, program management concepts, and group interaction psychology are typical of broadly applicable approaches. Of course, subjects such as physical measurement principles and equipment, manufacturing control concepts, supplier quality engineering and support, and the quality data and information program—which are associated primarily with just the Quality System—must also be included in the training program.

It is not to be inferred that all of this material should be covered in one session or even in one course. As was indicated above, these are suggested topics only; the actual material to be presented could be covered in a number of courses, if merited by circumstances. The purpose of such training—to provide the basis for thorough understanding, acceptance, and practice of pertinent Quality System and quality control programs and techniques—should be recognized early and retained, however. Only through such constructive use of the System approach can we achieve the quality, cost, and time improvements the Quality System can provide.

8.3 New Operator Quality Familiarization

In the overall area of personnel training for the Quality System, one vital portion is that covering production operators. Whether these are newly hired employees or experienced ones being transferred to different jobs, it is important that we provide the necessary training to enable them to produce a quality product consistently. Such training obviously includes providing them with the job skills they will need: tool and equipment handling and care, safety practices, team participation, and the use of production documentation.

Just as important are the quality aspects of their activities. They must be aware of the standards against which their work will be judged, the type of inspection their own and others' work will receive, and the basis for generation of quality data. They must have an understanding of how to handle defective material, of quality control techniques applied to their work by themselves and by others, of the effects of neglect or insufficient maintenance on the capability of their equipment to do the quality job, and of the impact of the quality of their output on later internal operations and on customer satisfaction.

The New Operator Quality Familiarization training program may take one or more of several different forms. Many organizations operate training or orientation centers for new employees. Some also use the same or similar facilities for update training of experienced operators and for retraining of transferred or reassigned operators. Other companies have the line supervisors, perhaps assisted by their own or support department personnel, conduct formal training for their operators within established time constraints with respect to job start dates.

On-the-job training, either by the supervisor or by other operators or indirect specialists, is used to some degree by virtually all organizations. Since this approach requires the minimum of facilities, resource application, and preparation, it is often the sole training method employed; it is also the most variable in effectiveness, the most subject to "sloughing off," and the least likely to provide proper quality indoctrination. For these reasons, knowledgeable managements always combine on-the-job training with other, more formal training programs.

During the training process, new employees especially should be continually apprised of the benefits to them, to the organization, and to the customer resulting from their production of a quality product. Their help in identifying the sources of and correcting quality problems should be solicited, and the supervisors, of course, should be oriented toward reacting favorably to such inputs from the operators.

With a properly trained work force that recognizes the cause-and-effect implications of their work and the utility of quality-control techniques, an organization will realize efficiencies in equipment, material, and labor utilization resulting in increased productivity and improved customer satisfaction with the product. Other benefits include both lowered inventories, with attendant carrying costs, and lowered employee turnover rates. Informed employees are more likely to be satisfied with their jobs than those who perceive that "the company" is not interested in their contributions.

8.4 Inspector and Tester Training

Regardless of organizational structure or titles, whoever performs appraisal work is an "inspector" in the sense of this element. To be a successful inspector requires certain abilities, skills, and knowledge, the specifics of which are determined by the conditions and items being inspected. Whenever an organization plans to assign some person to an appraisal activity, it becomes necessary to determine which of these needed tools might be missing. This determination is made in 8.11 Qualification Standards.

Once such lacks are identified, the organization must determine whether they can be overcome by training and, if so, what kind would be most effective. When training can remedy the situation, the organization will proceed to develop the necessary programs. Often common needs are apparent among several or many inspectors; this, of course, calls for group training, of both the formal classroom and the on-the-job varieties. In other cases, individual training will be required.

The key point is that the organization commits itself to provide whatever training is required for each inspector to be able to do the job effectively and efficiently. This will result in employee proficiency in applying appropriate Quality System techniques, in relating their work to the product in use, in performing its equipment-related portions, in generating the resultant data and useful information from their job activities, and in doing these things in a timely manner. The inspector can then achieve a certification level of capability, increasing his usefulness to the organization and making his work far more interesting.

Records of training and retraining, including those of certification achievement and subsequent requalification tests, should be maintained for each inspector. This practice is important not only for the morale of the involved personnel, but it can be useful also in dealing with product liability claims and regulatory matters and in conducting labor relations negotiations with unions and others.

When properly conducted, Inspector and Tester Training ensures the availability of the specialized and expensive "inspector" resource at optimum readiness for application.

8.5 Field Personnel Training

Both products and services involve their producers in contacts with their customers. Obviously, there are sales and advertising contacts; in this program of Field Personnel Training, however, we are concerned with application, erection, installation, and service activities.

Application relates to the custom engineering of the product, in concert with the customer, to suit the specific situation in which the product will be applied. Where such activity goes beyond selection of available options and actually affects the design or the performance potential of the product, proper controls must be applied.

Erection and installation are similar to each other, since they both involve placing the equipment into the customer's environment and getting it to work there. Erection typically involves physical construction—a power plant might be a good example. Installation can involve as little as delivery, plugging in, and turning on a TV set, or it can require extensive modification or rearrangement of facilities, such as putting a large numerically controlled machine tool into an existing factory operation. In either case, much of the product's capability to satisfy the customer is dependent upon the talents of the field personnel involved.

Very few products, other than consumables, never require service; and even those few products and consumables may generate complaints, which must be dealt with properly. Therefore, while application, erection, and installation may not apply in a given situation, service always does. It is vital to any producer, then, to ensure his service capability.

Organizations that control those personnel who may affect or contact the customer are obviously in a favored position to accomplish this, but all organizations need to find a way to train field personnel sufficiently. The training has to include product knowledge, safety practices, job skills, supply procedures, and customer relations. It should also incorporate quality subjects, such as inspection and test, reporting required information, troubleshooting, product liability implications of the work, and methods of obtaining any needed support from the producer.

The Field Personnel Training program must be an ongoing one. Its details must be tailored to suit the company's needs and those of the persons being trained, but the concept must be clearly understood: The customer must be involved with only those personnel who are adequately trained. When this is done, the company will have eliminated the major source of customer dissatisfaction with an otherwise reliable product.

8.6 Rotational Training

The development of a versatile work force is an advantage for most organizations. Employees who can handle assignments in several different areas are definite assets in cases of unexpected loss of key personnel and during periods of expansion or when broad-gauged management at any level is needed. Many organizations have carefully developed programs for deliberate rotational assignment of their personnel.

Within the Quality function, inspectors, testers, technicians, engineers, and managers should be involved in such a Rotational Training program, moving among the various activity groups on a planned basis to an extent compatible with their backgrounds and interests. In addition, at least some of these same people, particularly the managers and engineers, will benefit from assignments outside the Quality function, just as people from other disciplines will benefit from rotational assignments in the Quality organizations.

All such moves should be thoroughly planned for, settling details of duration, assigning specific responsibility and authority, providing sufficient introductory training to minimize operating and safety risks, and articulating expected benefits for the person and for the company. These plans should be communicated to everyone involved far in advance of the start of the assignment. Ordinarily the exposure to a new environment lasts six to twelve months. Evaluations of the experience and its results should be made by a third party at least once during and shortly after completion of each training period. Everyone should be encouraged to view the assignments as opportunities to grow, to contribute to the success of the organization, and to investigate what may turn out to be superior areas for permanent employment. As a minimum benefit, the new relationships developed will aid the individual in the successful performance of later duties.

Particularly in the inspector, tester, or technician level of such programs, individuals may not wish to participate or may have rare skills uniquely suited to one needed activity. In these special cases, their involvement in the program should be limited or eliminated. However, most people having such assignments will benefit from the program.

8.7 Participation in Company-Sponsored Training Programs

Basically two types of training are called for by the Quality System and provided by an organization: training the employee to do a job that the organization needs to have done, and training to prepare the individual for a different or larger role in company operations. The first is likely to be viewed as mandatory; the second is an option for the employee.

In addition, we can consider that the training may be direct or indirect. Direct training could involve achieving technical certification in a specialized field such as quality engineering or radiographic techniques. As examples of indirect training, we might consider taking part in professional society activities or serving on trade association committees concerned with quality questions. Even the indirect training programs may be of sufficient importance to the company that participation by selected individuals may be required.

The training programs may be conducted in-house by company or outside personnel or may consist of support for courses offered by academic institutions, consultants, or professional societies. Sometimes company personnel may have to become qualified as teachers to provide the potential for satisfactory instruction of others in the organization. When proper operation of the Quality System *requires* that key personnel display certain capabilities they do not currently possess, satisfactory completion of relevant training becomes a condition for continued employment. This is as true for engineers and managers as it is for shop operators, inspectors, and testers.

The company must determine its policy of financial and other support for both mandatory and optional training, ensuring equitable administration of the policy throughout the organization.

The important consideration about the concept of this element is that the employees of a company must provide the broad base of skills the company needs to meet the objectives of the Quality System. With properly identified and selected training, both immediate and future needs can be satisfied.

8.8 Training Effectiveness Measurement

Several elements of the Quality System specifically address various training subjects and audiences. Among these are 3.7 Quality Engineering and Training for Suppliers, 8.1 Overall Training in Quality System Subjects, 8.2 Management and Technical Personnel Training, 8.3 New Operator Quality Familiarization, 8.4 Inspector and Tester Training, 8.5 Field Personnel Training, 8.6 Rotational Training, 8.7 Participation in Company-Sponsored Training Programs, 8.11 Qualification Standards, and 8.14 Field Quality Orientation Program. Virtually all the other elements require some training to inform the involved people just what they should do and why. In the aggregate, the Quality System requires a great deal of planned, formal training of practically everyone in the company.

Such a substantial investment requires that the training programs be evaluated to assure that they are worthwhile so that the training will be applied for the maximum benefit of both the organization and the employee. While a number of plans have been made and studies conducted to measure training effectiveness, the effective ones seem to have all or most of the following characteristics:

- They set specific learning goals for the students
- They attempt to evaluate the training process while it is in progress, and at the end of the course, by third-party observation coupled with measurement of student learning
- They obtain an evaluation of the acceptability of the course content and presentation *from the students* as the course is being completed (or immediately thereafter) through questionnaires
- They try to assess the degree and accuracy of the students' application of the course content from the students and their supervisors at some time after course completion (often six months) and again perhaps a year later
- They use the instructor as a coach and evaluator, either on a fixed schedule after the course ends or "on call"

Of course, whenever any of these measurements result in an indication that improvement in the training effort is desirable, these organizations investigate the situation, determine whether the problem arises from the instruction or from the management circumstances within which the employees must work, and develop and implement appropriate corrective action.

As we apply the Training Effectiveness Measurement element effectively, we will see increased acceptance of our training programs, improved quality and productivity performance of our people, and reduced investment in marginal or useless training efforts.

8.9 Personnel Development

As part of the description of 8.7 Participation in Company-Sponsored Training Programs, it was pointed out that one of the two types of such training programs was oriented toward permitting the individual to prepare for a different or larger role in company operations. This is the concept generally understood when Personnel Development is mentioned. Assisting each person to achieve his full potential must be a major concern of Quality System managers; otherwise, the organization will not be efficiently utilizing its most limited resource—capable manpower.

To accomplish such an objective, each manager should describe to each employee just what is expected of him in the way of key activities to be performed, skills to be acquired, schedules to be met, and specific goals to be reached for full effectiveness on the job. Beyond that, each manager should counsel all employees, especially if asked, as to what additional education or training they would find most useful in preparing themselves to advance in their careers. This should be done on an individual basis; only a few aspects of such a program can be handled effectively for people in groups.

Since there are people who lack aspiration to supervisory and management positions, not everyone will benefit from an MBA. The manager must structure his advice in the context of what the employee wants to do as well as what his performance indicates he can do. Many an excellent engineer has been promoted to be a worthless manager—and both the new manager and the company have suffered as a result. We see a similar disturbing result from school and parent pressures. Children are pushed into college and are permanently "turned off," rarely becoming the excellent and happy automobile mechanics they otherwise could have become.

The manager, then, has a responsibility to determine, as best he can, just what each employee's capabilities are. Applying the methods described in 8.12 Quality System Personnel Performance, the manager will also use well-developed job descriptions—both for the employee's present job and for those most likely to be reasonably attainable by the employee—as the basis for identifying the additional training and experience the employee needs to qualify for an advancing career.

An interesting Personnel Development program involves rotational assignments in which the employee is moved from job to job, even from department to department, for exposure to a number of disciplines. The employee obtains a generalist's view of the business and thus becomes more valuable to the company. (See 8.6 Rotational Training.)

Both the employees and the company stand to benefit when employees at all levels become involved in the quality-related work of outside organizations. These can be community groups, professional societies, trade associations, and standards-writing and -administering bodies. The company should therefore encourage employees to participate in such activities and support them effectively as they do so.

When this entire program is effectively carried out, the manager, heading a department that has solid backup for each position, becomes more easily promotable. In addition, when people are advanced into other departments as the result of the manager's counseling and support, the net effect will be to broaden his overall influence within the organization.

8.10 Personnel Quality Participation

Quality of products and services is, directly or indirectly, almost always the result of what people do or don't do. Machines and materials are often involved, of course, but in the broadest sense their contribution to quality is heavily dependent on people. Designers, planners, supervisors, doers—they determine almost completely whether the equipment will do the job, whether the material will produce the desired end result, and whether the product will satisfy the customers' requirements.

The Quality System recognizes this dependence on people in a process that starts with the marketing personnel identifying the need for a new product or negotiating requirements with a potential customer. Personnel Quality Participation extends through research, design, development, purchasing, suppliers, receiving, facilities, manufacturing planning, the quality function, production, packing and shipping, and warehousing and distribution, to erection or installation and service. It involves inputs and participation from finance, legal, data processing, personnel functions, community relations, maintenance, materials, production control, and all the other people who contribute to and affect the organization's ultimate ability to satisfy its customers with the quality, reliability, and safety of its products and services and to satisfy its shareholders with the effectiveness of all its operations.

Most top management people have recognized that quality is people-dependent, but they have usually not realized the extent to which this is so, nor have they understood the system effect—the interdependence of everyone's actions, including their own. For this reason, most attempts to "motivate" people to "do a better job" have been directed only toward those who typically are least able to control the end result—the production operators. Such programs are doomed to failure despite initial successes, often with a net worsening of quality and costs, because the primary cause of quality problems is management failure. Whereas some authorities attribute 80 percent of such problems to management, my experience indicates that this figure is too kind to management. Whatever the proper allocation of blame may be, however, the production operator, often charged with correcting all the problems but unable to control more than a small fraction of the result, becomes frustrated. It is this frustration, with its resultant sense of failure and growing suspicion of management's motives, that produces the net worsening of quality and costs.

One program for involving the work force in identifying and correcting problems has been used for many years. It is commonly referred to as a suggestion system and usually is based on a concept of monetary rewards (akin to profit-sharing). The arbiters of such a system are usually some technical group, such as Industrial Engineering, that is often understaffed. As a result, many suggestion systems are woefully administered, with resultant negative effects on morale and the credibility of company management. For a suggestion system to be truly effective, resources must be put in place to ensure rapid processing of all entries—with a strong orientation to accept all suggestions rather than rejecting them.

The sense of this discussion, then, is that management must correct its own sins before, or at least simultaneously with, asking the operators to correct theirs. During the 1960s, there was a rash of packaged motivational programs, the most famous of which was the "Zero Defects" program of Martin-Orlando. In concept, this program recognized the broad scale of personnel involvement in quality. All employees of a division, including the general manager, and suppliers and field service personnel as well, were expected to sign a pledge that they would henceforth produce nothing but perfect output. In practice, what worked eminently well at Martin-Orlando had a mixed result elsewhere and, although adopted by the U.S. Department of Defense as a government and defense industry-wide program, has quietly died away except in a very few instances. It has, however, a very wide application in Japan, for reasons discussed in 6.7 Quality Level Improvement.

Why the mixed results and gradual disenchantment in this country with a program with such potential? Most companies who adopted the program did so without understanding the necessary preconditions for success. At Martin-Orlando the program was developed to attack the few remaining direct shop operator sources of problems *after* management had eliminated the great bulk of its contributions to bad quality through the application of sound statistical quality-control techniques coupled with imaginative and effective corrective action. This means that the efforts of individuals to improve the results of their operations *could* be effective. In addition, Martin-Orlando had trained its people thoroughly, had provided them with proper work instructions and visual aids, and had installed or was prepared to install equipment capable of producing the required quality of product. The company had also promoted and followed the same types of activity among its suppliers. Very few American companies went this far or saw the need to, and as a result the majority failed to realize the benefits of their programs.

Defining a "Good" Job

What is required to motivate a work force to produce a quality output? With vanishingly few exceptions, everyone would rather do a good job than a bad one. Unfortunately, defining "good" and "bad" is extremely complex. If a design engineer has a release deadline and meets it, that's a good job. But what if the design is capable of being produced only with considerable nonconformance using state-of-the-art manufacturing techniques? What if the design contains latent field failures or personnel hazards when used in a way somewhat different from what the designer anticipated? What if aggregate service costs to the user exceed the purchase price in three years? Was it such a good job to meet the release date?

If a production foreman meets his production schedule for the month—that, too, is a good job. What if ______? (You fill in the blank.) The point is that "good" and "bad" depend on management definition and emphasis. Like it or not, people tend to do what they think the boss wants them to do, and they obtain their interpretation of this from many sources. They hear what he says, both positively and negatively; they hear what he doesn't say; they see what he does and what he lets them do; they react to his manner in general and as it relates directly to them. They view him as management, as "the company," and they accept his views of what is good and bad.

So the answer to what will properly motivate the work force is to define "good" and "bad" clearly and unequivocally for every employee—from the chief executive officer to the newest janitor, and for supplier and field personnel as well—and then to train all levels of personnel to try to achieve what is good all the time. Not because they won't try to achieve it, but because it must be visible at all times that *good* is what the boss really wants. This means that both capital and expense money and personnel will be made available to correct previous management errors and to prevent new ones. It means that successfully doing the good things will be rewarded by all methods available to management instead of the previous practice of rewarding the short-term or limited-scope good, which often have bad later effects, or even sometimes of outright rewarding the bad. And finally it means that these "positive reinforcement" devices will be made continually visible to all involved personnel so as to establish firmly and permanently the conviction that the boss wants good results. In this endeavor it is unquestionably true that "Once Is Not Enough!" and, to adapt slightly, "Words Are Not Enough!" There must be solid, regularly reiterated deeds, demonstrating this management commitment to all.

With such a climate, specific programs to enhance understanding and involvement can be effective. Experience has shown that sustained programs eventually lose appeal; the hit-and-run approach introduced after the first phase has operated for several months or a year is better suited to achieve long-term results with a reasonably stable work force. Proper planning for and bringing new personnel rapidly up to an involvement level comparable to that of older employees is also necessary.

The Quality Circle Approach

Another type of program to achieve and sustain involvement has been to turn over to the line personnel the responsibility for eliminating the sources of quality problems. In Japan, a sizeable portion of industry and industrial workers is involved in the Quality Circle approach to problem solving. The Japanese culture makes this technique especially effective, but with adaptation it is applicable in any culture.

Fundamentally, the Quality Circle concept gives teams of production operators, supported by supervision and technical personnel as required, the authority to identify problems that need to be solved, to investigate and determine the cause(s) of the problem, to develop and test solutions, and to apply the successful solutions. All this is done in a climate of constructive competition, with company, city, region, and national encouragement and recognition of accomplishment. Most of the problems solved this way are of minor import, but the aggregate is large; and the side effects of an increased sense of community and commitment are enormous and probably of more benefit to all than the actual effect of the solution of the problems.

A related but somewhat different approach is to form multifunctional teams to address assigned problems involving more than one department. Such teams can be established at any level, from top management to operator, be given one problem or many, and have a fixed or indefinite life. Most organizations using this technique usually restrict any team from comprising more than two management levels and only one when the team's membership is made up primarily of nonmanagement personnel.

Some companies call such teams "Quality Circles" also, but other names are more common. The methods of operation tend to be similar to those of "Quality Circles," but the problems addressed and solved are usually thought to be more critical to the company's success. It is appropriate to use both the "Quality Circle" and multifunctional team approach in an organization.

The adaptations necessary to make concepts like these work well in other societies—ours, for instance—involve recognizing that the typical employee may have paramount interests *outside* his place of employment and that most companies will not already have applied as fully, or have the management understanding of, the quality-control techniques already used routinely in Japanese industry. Essentially, the preconditions are similar to those for Zero Defects—management knowledge and commitment and the prior elimination of the worst of the bad conditions for which management is responsible. Also required is thorough training of the participants in the use of the team-building and problem-solving techniques and the presentation methods needed to be effective in such an activity. (See 8.2 Management and Technical Personnel Training.)

A necessary companion to understanding and involvement is interest. Whereas involvement implies an identification by the employee with the objectives of the company, interest implies his recognition of and concern for his own job. In recent years much has been said about worker dissatisfaction and boredom on the job, particularly addressed to repetitive manual and clerical labor of the assembly-line type.

Experiments have been made in the area of job enrichment, giving the employee responsibility for more or perhaps all of the job. A notable example is the Volvo program in which some assembly lines have given way to stations where a team assembles an entire car. Some employees thrive in such an environment; others do not. The determination of what is needed to combat disinterest effectively in a given circumstance should be developed jointly by management and the workers involved. Rapid turnover and employee unrest of any type are signals of disinterest and should not be ignored. In many cases, however, changes in work practices are not required and would be unwelcome to affected employees. The jobs as they currently are, plus training and other System activities, may be sufficient to alleviate whatever boredom may exist. Develop major programs in the interest area only after assuring yourself that they are needed, and then pursue them vigorously.

Involve your employees in the long-term success of your quality-improvement efforts by obtaining feedback from them on all conditions affecting their work. Surveys, suggestion systems, and "rap" sessions are mechanisms frequently used for this purpose.

When participation programs are properly developed and dovetailed with the training and personnel development elements of the Quality System, and when the employee detecting or foreseeing a quality problem is given a mechanism for resolving it whether or not his supervisor concurs, they will identify and overcome areas of weakness that now exist in motivating all employees to do a quality job—weaknesses in management commitment, training, communications, and positive involvement by all in quality-related activities. The net effect will be significant improvement in quality, costs, and operating times.

8.11 Qualification Standards

Recent developments in the United States have rendered suspect the application of qualification testing for job applicants. The underlying consideration is that disqualification as the result of such testing may reflect a cultural bias rather than an applicant's genuine inability to perform. The use of the same or similar tests solely to determine what training must be given to the applicant to develop full job capability—with the training then provided by the company—is fully acceptable, however.

Although this process may take more time and result in more marginally effective employees, at least initially, the net result should be the same: a body of employees capable of performing all aspects of their jobs to meet acceptable standards. In achieving this end, a company will see at least two parts to the action of obtaining a qualified work force. The first is to determine what the employee would need to know and do to be able to perform his work satisfactorily. Often these criteria can be broken into two parts: what all employees require in common and the unique requirements for a specific assignment. When the latter set of requirements is sufficiently demanding, the job may warrant skill certification, by the company or by an outside agency. There are internally certified machinists, welders, solderers, testers, and so on. The same skills may be externally certified, as may be quality technicians, quality engineers, reliability engineers, and nondestructive testing technicians (under the certification programs of the American Society for Quality Control and of the American Society for Nondestructive Testing), or personnel exhibiting other capabilities. (See 4.3 Special Process Control.)

The second part of the process of building the proper work force is selection and training. While the use of testing for selection purposes must now provably reflect genuine job requirements without cultural bias, it can and should be used for those purposes when warranted. A visual inspection job that truly requires stereoscopic vision cannot be filled by a one-eyed man, nor can a job requiring the application of abstruse mathematical concepts be expected to be performed by a high school dropout. However, tests to detect less blatant misfits frequently show remediable defects in the applicant's skills. This serves as the basis for selective training that, coupled with normal orientation programs, will provide the applicant with the necessary tools for successful job performance. To accomplish this requires the organization to dedicate sufficient resources to the development and implementation of both general and tailored training programs.

This activity is particularly important for those personnel typically hired in large numbers, such as production operators, material handlers, inspectors, testers, technicians, installers, servicemen, draftsmen, and, in some situations, engineers. It can be usefully applied even to lower levels of supervision and to sales personnel. Some companies, with measurable benefit, apply the concept to all new and transferred employees regardless of level. Essentially, what we are discussing is the idea of determining carefully what the person needs to know to do a job properly and then making this knowledge available. Proper standards, measurement of attained skill levels, and provision for indicated training are the ingredients of a successful program.

Two further factors complicate the process of obtaining the benefits from having a qualified work force: attitude and interest. A negative attitude on the part of a qualified employee typically damages and may even destroy his ability to do the job. A positive program of selection and training will help to prevent a negative attitude from forming but is unlikely to be wholly successful without the program of 8.10 Personnel Quality Participation. Interest and its lack are discussed in the same element. Disinterest can result in nonconformities being produced through induced carelessness, even when such result may be emotionally disturbing to the person responsible.

Once qualification standards are established, performance standards can be set for the actions of Quality System personnel. The use of such standards for performance evaluation is discussed under 8. 12 Quality System Personnel Performance. A further outgrowth of the development of these standards is the resultant ability of management to determine personnel needs consistent with production, service, and other requirements. The use of historical ratios to establish manning levels of appraisal personnel without considering the requirements of a new, much more sophisticated product, for example, is an exercise in poor judgment. Proper application of the knowledge derived from utilization of this program will prevent such errors.

With Qualification Standards properly developed and used and with identified training and personnel development needs satisfied, we can expect to have a work force both capable of and interested in meeting the company's overall objectives.

8.12 Quality System Personnel Performance

As individuals and as departments, it is necessary that all personnel involved in the Quality System perform their work in such a way as to achieve the Quality Objectives. This implies that their roles have been fully defined, that they have been properly informed and understand what is expected, and that specific goals and accomplishment schedules have been established and agreed to by all as described in 1.15 Quality Program Effectiveness Review. This element requires the setting of these goals and schedules in all the control measurement areas of the Quality System, the preparation by each department of its plans to achieve the scheduled results, and the actual tracking of results with indicated corrective action.

Although departments may be able to produce measurable impact in one or more of the six Quality System control measurements, individuals and small groups typically cannot, and their performance is usually not measured against those yardsticks. For them, performance is normally judged with respect to the specific requirements of their jobs: conformance to procedures, reaction time and effectiveness, productivity, quality of output, and other pertinent criteria.

For example, inspectors are often measured by their efficiency; i.e., the percent of their decisions (both "accept" and "reject") that are correct. Careful tests indicate that, for experienced personnel working under good shop conditions, efficiency ranges from 65 percent to 98 percent—with an average of around 80 percent. Where equipment is used that is only minimally affected by the inspector, the error rate is substantially reduced. However, because of the above "real-world" condition, an important step is to include in the calculations of 6.7 Machine and Process Capability a determination of the "measurement error" as part of the total variability.

How to measure all of the different pertinent aspects of personnel performance efficiently is difficult to determine. Techniques used are audit, examination of production and other records, subject and skill testing, interviewing, using accounting data, and many other approaches designed for the specific criteria selected as measurement bases. Whatever is selected should be a valid measure, yet at the same time be economical to operate. Where practicable, the person(s) to be measured should contribute to the development of the measurement.

It requires considerable ingenuity to accomplish all this measurement, compare the data with the established standards, and feed back discrepancies for investigation and correction while still minimizing costs. The data program should make use of existing data sources—production data, for example—and be joined with other Quality System data needs to reduce requirements for additional unique data processing capacity.

In all of the reports measuring performance, trends should be shown so the effects with time of preventive and corrective programs can be seen. Such an analysis can guide the selection of the proper type of corrective action program (e.g., training) that might not otherwise be obvious. It is implicit that excellent performance by individuals or groups should be suitably recognized by management. Public recognition of the accomplishment—based on fact, not fancy—will be accepted by the entire organization as evidence of management's true interest in quality. Monetary or merchandise awards can be counterproductive. Best results seem to be obtained by visible indicators of sincere appreciation for excellent results, such as publicity or reserved parking spaces. A mechanism should be developed and applied to relate the recognition processes used to resultant changes in morale and employee motivation for quality and quality improvement.

The Quality System Personnel Performance program must cover all departments and all levels of personnel involved in the Quality System, from operators and inspectors to department heads and beyond. If the individual, group, and departmental goals are properly established, circulated to, and accepted by all concerned and programs are established and operated to accomplish them, the Quality System Objectives can be met. Without such a program, successful attainment and maintenance of the Objectives will be very unlikely.

8.13 Quality Information Transfer

The deliberate and formal transmission of necessary and useful information about quality and quality control to all involved users of such information is covered in Quality Information Transfer. This transfer of knowledge takes place in essentially two ways. The first is through carefully planned meetings with selected personnel involved in the Quality System where quality problems, practices, and programs are presented, discussed, and acted on. The second is through bulletins, newsletters, posters, or other communications mechanisms by which the same type of information may be transmitted to an even broader audience.

Each of these vehicles should be used only when there is something worthwhile to be presented. Holding a meeting or issuing a bulletin every week just to adhere to a pattern defeats the purpose; people "turn off their hearing aids" when there is nothing useful to be heard.

Meetings and documents should be brief and informative. If the discussion in a meeting is becoming repetitive or inconclusive, the chairman should end it. The important thing is to ensure that every meeting produces a planned-for and useful result: a technique learned, a problem solved or organized for solution, or a program understood, perhaps through development in a series of meetings, and accepted. As much as is practicable, the document effects should also be measured and required to produce a similar desirable result.

From each exercise of this element should come an increased appreciation of Quality System concepts and an application of "do's" and "do not's" to a broad spectrum of situations, not just that of the instant example. It is also sometimes useful to send certain types of written communications to each employee's home. Family and friends may benefit from such information and identify better with the employee's work situation.

When these types of communications are properly used, the organization sees a reduction in wasted time, more rapid resolution of problems, and a readier acceptance of the Quality System programs because the people involved understand them better.

In addition to the necessary communication of Quality matters to the internal audience, it is important that the organization provide related information to outside groups as well. This should involve all levels of management, including the top, and professional employees. The effect on the community, trade associations, professional societies, suppliers, and customer groups can be positive only when sound "Quality" positions are taken on improvement in product or service, plant safety, employee relations, environmental controls, health care, and business ethics.

8.14 Field Quality Orientation Program

Once an enterprise has developed and installed a Quality System, it becomes desirable to capitalize on it competitively. Since one major result of applying the System is improved quality, reliability, and safety of the product and service provided to customers, that fact and its implications should be made clear to all of them. When a manufacturer is a supplier to other industry or commercial customers, mutually beneficial programs can often be established as a result of the supplier's new approach to quality. (See 1.11 Product Certification and 3.6 Supplier Certification and Objective Evidence of Quality.)

The necessary information can be provided to customers in numerous ways. One method that is very effective, since it serves an internal purpose as well, is to provide a formal Quality System orientation for all company personnel who are expected to be in direct contact with the customer. This orientation can either coincide with or precede a similar program for the customer's personnel. It should also be extended to stockholders and the community in general, perhaps somewhat modified, to improve public image and to promote the use of beneficial techniques throughout the area. Company personnel involved would include those from the sales, distributor, dealer, installation, and service organizations. In addition, marketing and advertising personnel would receive the same orientation for use in their key activities of customer motivation and research.

Some of the subjects to be covered in the orientation program would be:

- Awareness of the critical nature of Quality for organizational success
- The scope of the Quality System
- An overview of the new product design control activities
- A review of the factory-related controls
- The interface between field and factory, emphasizing the resulting delivery improvement and the importance of timely feedback of pertinent and accurate field performance and failure data (see 5.5 Field Quality Data Reporting Control)
- The impact of the System on 1.8 Product Liability Prevention (this section may be excluded in the program for customers)

The formal presentations should be tailored to the needs and interests of those personnel actually to be included in the audience. Factory tours and demonstrations for some, videotaped shows for others, and visits by "flying squads" for still others are only a few of the approaches that have proved useful.

The Field Quality Orientation Program will improve the field personnel's level of understanding of the Quality System and its expected effects, thus providing a basis for their rapid alerting of the factory should some unexpected development occur in product use or service. They will also be better able to present the company's products and services favorably to the customer.

Subsystem 9

Customer Contact

This subsystem, Customer Contact, is oriented toward obtaining sufficient control over field activities to ensure that the customers' needs are fully satisfied in such a manner as to produce extensive repeat business. All of the activities of this subsystem may not pertain to a given company because of its product (e.g., installation, product service, and several other elements are not required for foods), but, with that reservation, they all apply, regardless of the size of the company.

In many places in this book, reference is made to customer satisfaction as a critical measure of the success of the enterprise. Virtually every element of the Quality System ultimately has an impact on customers, but this subsystem contains those that are normally most highly visible to them and with which they deal directly.

The customer evaluates any enterprise from as many as four or even more viewpoints: the sales contact; the product, including purchase price, delivery, ease of placing it into use, operating cost, and reliability; service availability, effectiveness, and cost; and the public or community image of the company. The factory and all the great things that are done to make it efficient and safe are essentially invisible to the customer—except as an indistinct part of the community image or as exposed through advertising, news media, or other contributions to the company's public image. This subsystem, with the help of a few elements in other subsystems, covers activities oriented to meeting all of the customer satisfaction criteria that are visible to the customer and are logically part of the Quality System. With the exception of the sales contact factor, all of the criteria are either controlled or contributed to by the elements mentioned.

It may be helpful to introduce certain of the elements of this subsystem early in the process of developing a Quality System. This is true even though the major effects in the field result from the preventive and corrective measures taken before the product is ever shipped to a customer. When the organization needs rapid evidence that the Quality System will make an important contribution to the company's success, however, improving the customer's quality experience can be dramatic and convincing. Among the elements that could produce such a result in certain circumstances are 9.2 Installation and Field Test, 9.9 Customer Data and Support Program, 9.10 Field Problem Predictions, 9.11 Installation and Service Difficulties, 9.12 Field Problem Controls, 9.13 Field Problem Handling, and 9.15 Audit of Field Quality Activities.

The elements of this subsystem are arranged in the order they would likely be put into use by a new enterprise.

Note: It has recently become fashionable to apply the word "customer" to a colleague or other department of the same organization who is the initiator and the direct or indirect recipient of the result of some activity of the Quality system. While such identification is legitimate, and while it is necessary to adopt that philosophy in the workplace, the indiscriminate or broad use of the word can be confusing. To minimize the problem, some organizations identify the different classes of customers by phrases such as "internal customers," "external customers," and "government customers," among others. There is a third class of customers in some industries—dealers and distributors—and even a fourth (regulatory agencies who are not actually purchasers of the products involved, except perhaps for testing purposes) in others.

The key to all this is to establish an organizational policy on this subject, with definitions, and then to ensure that everyone is informed and follows that policy. Having done that, you will have little difficulty adapting the material in this subsystem appropriately to whatever class of "customer" may be appropriate for your purposes.

9.1 User's Safety Warnings

The entry of the U.S. government into broad-scale regulation of many products, coupled with response to the pressures associated with product liability, has led to major changes in many manufacturers' approaches to dealing with their customers and with field personnel. It has become both mandatory and politic to inform product purchasers, installers, users, and service personnel about any hazards associated with the use, likely misuse, or service of a product. The hazards may be direct or indirect, proven or suspected, limited or general, certain or possible, but notice must be given.

This notice should be clear, complete, preferably concise, and usually permanent. It must be displayed where the user can hardly fail to see it; thus the owner's manual is a less desirable location than one on the product itself. Sometimes the best place for the notice is on the outside of the package, particularly for nonreusable items, or where it becomes visible just as the package is opened. Placards, decals, tags, plates, stampings or embossings, and labels of various types are used. Multiple language considerations, including Braille, are often important.

The messages of the notices should be tested and the proper level of understanding by users verified before publication begins. A high level of communication efficiency, which cannot be assumed without test, is required for this information.

One of the benefits of this User's Safety Warnings program to the manufacturer and to the customer is that thorough safety evaluation of the product design (see 6.3 Design and Analysis of Reliability and Safety Studies) will often cause the designer to eliminate the safety hazard rather than announce it and bear the associated costs and risks. Another benefit is the diminution of risk to both user and manufacturer through the user's recognition and avoidance of an otherwise perhaps unanticipated hazard. This is, of course, the primary purpose of the program.

9.2 Installation and Field Test

Many products, such as a toothbrush or an electric toaster, are ready for use when received by the customer; unpacking from a package may be the sole step required. Other products require varying degrees of sophistication and effort in getting them into operation—an under-the-counter dishwasher or a new home furnace, for example. For still others, the installation effort may be more extensive than that of the manufacturer who produced the items, such as elevators in tall buildings and power-generating facilities.

For a small appliance, there is no need for a separate consideration of installation quality. At most, a simple statement or two in an owner's manual is all that is required. In the other two situations, however, significant effort must be expended to ensure that the installation proceeds efficiently, safely, and with ultimate customer satisfaction.

To this end, it is necessary to produce the pertinent instruction documents, conduct the required training for installation or erection personnel, and provide necessary tools and equipment to control and check the work being performed. The training element is covered in 8.5 Field Personnel Training; tooling and equipment are covered in 7.1 Quality Measurement and Control Equipment Plan; this element, Installation and Field Test, is concerned with the instruction documentation.

The documents should cover received product quality verification at the installation site, the installation effort with associated in-process inspections and test, final readiness adjustments and tests, and the ultimate acceptance demonstration or testing for the customer. Provision also should be made for generating the necessary data for product design and manufacturing corrections, when required. These data will support the work of Quality System Personnel Performance (8.12) and Installation and Service Difficulties (9.11).

As we generate the documents to serve the purposes of this element, it is vital to consider the cultural level of the field personnel who will be using them. Clear, explicit instructions, with substantial use of photographs or exploded drawings, are normally far superior to wordy descriptions. We must be sure that safety precautions are properly emphasized and that quality and other performance standards are stated unequivocally. When changes in the product, equipment, required practices, or standards are made, the field documents must be immediately updated and obsolete ones removed from circulation, as covered by 1.5 Documentation Control.

When proper planning and documentation are accomplished, installation, erection, and field tests can proceed with maximum likelihood of achieving customer satisfaction with both schedule and quality while at the same time meeting cost goals.

9.3 Product Service

A manufacturer of a durable product that requires service in preparation for use or after the customer places it in use must determine how such a service may best be provided, especially from the customer's viewpoint. After making that determination (and after developing, testing, and implementing a service plan—with standards) the service must be provided rapidly, efficiently, and economically, albeit profitably. If the effort is to be provided by others, the manufacturer must ensure that it is supported properly.

In any case, such products are usually sold with a warranty or guarantee. These statements form a contract between the manufacturer and the user of a product and must be couched in clear language, cover all pertinent points that may be at issue, specify any limitations of coverage, and reflect the company's knowledge of the safety, durability, and performance capability of its products—as well as being fair to the customer. Of course, competitive warranties should be equaled or, if possible, bettered—with believable claims to support such positions.

Many large companies operate in more than one service mode. They may maintain a number of company-owned service centers or perform some service in-house to enable them to keep a finger on the pulse of the field experience and to try out new service concepts or programs to meet or exceed competitive offerings without premature outside exposure. At the same time, the bulk of their products and customers may well be served by licensed or independent service agencies or even by the customers themselves.

With all of these possible alternative approaches and combinations, over the existence of some of which the manufacturer may have little control, it is important to retain sight of the primary objective of Product Service: to achieve continuing customer satisfaction with the product and its producer. Customer satisfaction is obtained through initial and continuing contacts and impressions, price, delivery, easy access by the customer, performance, dependability, rapid response, operating and related costs, and availability. Efficient service produces and supports customer satisfaction wholly or partly in all these areas except initial contacts and impressions and price.

In each of these areas, the quality of the service effort has a major impact. Part of the service quality capability has to do with product design for service (see 2.9 Product Design Review), the parts (see 9.8 Renewal Parts Control), and the equipment (see 7. 1 Quality Measurement and Control Equipment Plan) used. But a major consideration has to be the people involved in Product Service and their abilities to interact positively with customers, take innovative measures when necessary (but ensure that these measures are technically sound and that their management is properly informed), and perform the service work optimally. This involves providing for their training, often in company-sponsored or company-operated schools. Rapid and effective support also is necessary from all company functions that they may call upon and for the availability of needed information on a continuing basis.

In the case of information, a fine line exists between too little and too much. If there are numerous sizeable service manuals supplemented by many bulletins and work instructions, the sheer volume may discourage the serviceman from using any of it. On the other hand, insufficient information may force him to improvise, with similarly disastrous results. In either case, customer satisfaction is endangered, sometimes indirectly, as in losing configuration control through unauthorized modification or incomplete service.

Recent advances in computer and display capabilities have led to portable service information terminals that are either fully self-contained or may be connected by telephone to a central data bank. The serviceman, almost regardless of location, then has ready access to all necessary information, with minimal logistical problems. A variation on this theme for permanent service installations is a combination information display and sequential test facility that provides the necessary connections and meters, with associated diagnostic equipment, in a fixed test stand. This integrates the equipment and information needs very efficiently.

One important consideration in the generation of service information is the educational and interest level of the potential user. U.S. Army automotive service documentation has been written in comic book style, profusely illustrated in color, and featuring "Connie Rod," a highly knowledgeable blonde mechanic. The approach has proven quite effective in accomplishing the service objective.

In any event, directions should be clear, brief, up to date, and contain all necessary information, including safety precautions to be observed by the serviceman.

For many businesses, properly managed Product Service is their most effective device for obtaining and maintaining continuing customer goodwill and repeat business.

9.4 Field and Factory Standards Coordination

When products require field erection, installation, or service for customer use and continued satisfaction, opportunities frequently come up for errors to be made and for disagreements to arise. Whenever this happens, there are resultant delays or later customer complaints. Many times these problems occur because the field people do not have the same kind of equipment or performance standards as used in the factory for control of production. In addition, the equipment they do have may be biased compared to that of the factory or may not be as well calibrated and maintained, and their standards may not be compatible with those of the factory.

The difficulties will be more or less severe dependent on the product type, complexity, and sensitivity to equipment differences, the expertise of the field personnel in dealing with such problems, and the ability of the factory to react rapidly and effectively when they become involved. Since most of these problems can be prevented, however, it is worthwhile to attempt to avoid them in any case, through Field and Factory Standards Coordination.

The factory and field organizations should cooperate to ensure the compatibility, if not the duplication, of their respective practices and equipment. Proper calibration and maintenance of field equipment must also be arranged for, preferably as an integral part of the Equipment Calibration and Maintenance (7.6) program. Field practices and standards for quality should be regularly checked against the factory's to ensure that they do not result in damage or in a deterioration of the quality, reliability, or safety of the product as the result of installation or service.

It is also important that each group be kept informed of the others' new or changed working conditions, practices, and equipment so that appropriate steps can be taken to maintain compatibility. Effort should be made to minimize duplication of assembly and test activities between factory and field. Cases exist where the factory performs an assembly and ships the result to the field, where it must be partly disassembled for installation; investigation reveals that there is no need for this extra work, that the factory can ship the product readily in the condition needed for installation.

Proper coordination of field and factory practices, equipment, and standards can result only in improved quality, cost, and time for field activities and in better satisfied customers.

9.5 Field Document Relevance

A manufacturer who produces a durable product requiring installation, erection, service, or operating knowledge must provide appropriate documents or other forms of instruction to customers and field personnel, whether company employees or independents. These instructions take varied forms, such as manuals, bulletins, cassette tapes, videotapes, movies, computer printouts or displays, labels, diagrams, charts, pamphlets, letters, placards, and even advertising. Much effort is, and should be, directed to making these items clear, concise, and complete, yet understandable at the appropriate educational level of the typical recipient and economical to use.

Although much has been done in this field, much still remains to be done. Some companies that, on an individual document basis, turn out superior material are confronted with limited field usage and numerous service difficulties; they have issued too many documents because of a varied product line, and no one reads any of them. In such circumstances, a completely new approach needs to be taken—perhaps tapes; perhaps computer diagnostic and problem response displays; perhaps a new approach to product design, eliminating product variation as viewed by the service arm; or perhaps a new educational approach to certification of service personnel.

Whatever the appropriate solution, the important thing to recognize is the changing structure of the marketplace and of the service community. This dynamic situation requires a correspondingly dynamic responsiveness by the manufacturer. Continual testing of the utility and effectiveness of the current approach is the watchword. The saying, "If we are doing it the same way we were a year ago, it's obsolete!" is nowhere more true than in the field. Yet that is often the last place many companies look to apply innovative concepts.

The owner's manuals we provide often read as though they were written forty years ago, and service manuals are often worse. With our advanced knowledge in communications, with our developed capability for producing exploded drawings and eye-catching illustrations, even with our recognition of the impact of comic book directness, it is remarkable that today's key documents often ignore all of these techniques and are just as lively and informative as a World War II Army Training Manual.

What we do produce should be reviewed and tested for accuracy, adequacy, and clarity; there should be no basis in these items for a successful product liability or personal injury claim from anyone attempting to follow our instructions. Only in this way can we be confident that we are actually satisfying the need for the proper information delivered to and applied by field personnel and customers—to our ultimate benefit.

9.6 Customer-Furnished Material

Suppliers to government agencies or to many large manufacturers frequently provide only a portion of the product they ship to such customers, and the customer provides the remainder. In other cases, the supplier provides a specialized service—plating, for example—and the customer submits partly finished items to be subjected to such service operation.

In either type of situation, the supplier, while holding the customer's property, has the responsibility to exercise stringent, accountable control over it. Often such material is rare or costly, and pilferage must be guarded against. Questions of quality frequently are of critical importance, particularly if the supplier has a contractual responsibility for the quality of the finished product. Although some customers insist that the supplier must use what is furnished to him, implying or stating that there is no need for incoming inspection or test of this material, it is often a matter of self-preservation for the supplier to ascertain the satisfactory quality level of the furnished items before assuming responsibility for them.

The critical aspects of this element, then, are as follows:

Assurance of the quality of the material as received

Stringent control over receiving, handling, storage, processes, and assembly operations involving this material—perhaps even at a more exacting level than for the supplier's own purchased material

Thorough accounting for scrap and other losses of the customer's material—again, perhaps more accurately than for the supplier's own material

Similar concerns and levels of control are involved when the customer furnishes manufacturing or appraisal equipment to be used to produce, inspect, or test product. Here there may be questions of equipment-use exclusivity records, special certification of or arrangements for its maintenance or calibration, and unique identification and accounting requirements for it—including when it is finally disposed of.

Because of the special nature of Customer-Furnished Material, including equipment, all the conditions or developments likely to be associated with it should be covered as part of the involved contracts. Such details as disposition of items found to be nonconforming before processing, scrap, other losses, obsolescence, use for noncontract purposes, purchase options available to the supplier, and wear-out are among the subjects that could profitably be covered.

By taking the steps indicated to ensure control of Customer-Furnished Material, the supplier can minimize the risks associated with the operational use of such items—risks that some suppliers, particularly those who sell to the military, have found to be considerable.

9.7 Drop-Shipped Item Control

The problem being dealt with by Drop-Shipped Item Control takes different forms with different product situations. Variations on the theme include items going to erection or installation sites, parts going directly to distribution warehouses or to service agencies, replacement items ordered directly, items in service stocks, and kit components and spares accompanying product shipments. The basic concern, regardless of the form of the problem, is how to ensure the quality of items shipped directly from suppliers to customers when used by the customer or servicemen. This concern exists because, in the customer's view, the product manufacturer is responsible for all his problems; the part supplier's involvement frequently goes completely unrecognized.

In any drop-ship case, the possibility exists that unsatisfactory quality product may be in the supply channel because it is not subject to the types of control exercised over material used in in-house production. One solution to the problem may be to have all such material sent to a central site where it can be examined as though for production. Another approach is source inspection and acceptance. Still another involves supplier surveillance and certification. A fourth calls for hardware audits to be performed throughout the distribution flow, including (for deterioration, damage, and obsolescence) in dealer and service stocks.

An allied problem, which the audit program helps address, is the practice by some service agencies of substituting lower-quality, lower-cost replacement parts for qualified ones.

In a given situation, combinations of all or some of these four approaches may prove most effective, perhaps coupled with dated stock, stock return allowances, and other service quality incentive programs. Whenever the programs fail and installers, service agencies, or customers identify nonconforming drop-shipped items, or when the controls uncover an unsatisfactory level of quality anywhere in the supply channel, the company must react positively and rapidly in correcting the problem. Such reaction may be complicated by questions of agency or of ownership of material, any work that may have been done on or with the item, and the numbers and geographic dispersion of involved items. Nonetheless, the company's reputation may be at stake, as well as the existing potential for product liability claims and, in some product situations in the United States, action by the Consumer Product Safety Commission.

Proper control of drop-shipped items can reduce customer complaints dramatically.

9.8 Renewal Parts Control

Mechanically, electrically, or even chemically active products intended for repeated use wear out or use up their active ingredients in some fashion. Sometimes they are worn out in long-term storage. The alternative to complete replacement of the product is to replace the worn-out or depleted part(s). Manufacturers of these types of products must plan for such replacement; in fact, it may be the most lucrative portion of their businesses. The replacement may take place at the customer's location, at a service center, or back in the factory.

Where the replacement is done, by whom, and under what circumstances is of considerable concern to those working with this Renewal Parts Control element of the Quality System because of the varying extent of control required in different situations. For example, if the replacement is made at the factory by expert repairmen, the part used may well be drawn from current production stores and be subject to no unusual controls at all. On the other hand, if a customer or serviceman is likely to be making a replacement in a worn assembly with a part taken from slow-moving stock in the field, the controls on the replacement stock could properly be much more stringent than those on the same item in production.

Parts intended to be placed in a worn assembly, for example, could require different tolerances, even different shapes, from the original production parts. Alternatively, renewal parts could require special packaging, preservation, storage, or shelf-life control instead of the usual procedure for their production counterparts.

Renewal parts, therefore, should be evaluated from a quality program planning viewpoint as being fundamentally different from the comparable production part. Whereas they may be provided from production lots of parts, they also may be sent directly from the supplier to the field. (See 9.7 Drop-Shipped Item Control.)

Once these parts are received in the field, the associated quality program must be concerned with achieving the basic objective that the part will be of satisfactory quality when installed in the product. This implies design provision for proper, nondamaging installation in the field; packaging and preservation for prevention of damage and deterioration in handling and in storage prior to installation; and stock rotation to minimize shelf-life problems in field stores.

As time passes, patterns of part usage and other factors may change, requiring a revision of the quality program to maintain or improve its effectiveness. It should therefore be reviewed regularly to ensure currency. This review should also touch on inventory control and stock ordering practices with respect to changing deterioration and damage experience and related Appraisal and Failure costs. (See 1.7 Quality Cost Program.) It may also involve modification of a JIT program in use for production.

The objective of this element is to produce an as-used service parts quality of a level to ensure that related quality costs are low, with no reduction in customer satisfaction.

9.9 Customer Data and Support Program

Many manufacturers find that they cannot serve their customers satisfactorily merely by providing them with good product at a fair price. Often dealings with customers involve records, certificates, and other documentation either generated to provide the supplier some competitive advantage or specified by contract or government regulation. In addition, customers or government agencies often provide resident or transient personnel who have legal access to the company's facilities, products, equipment, personnel, and records.

When such situations exist, a major aspect of the requirement normally involves quality. The military and space agencies and their prime contractors frequently assign inspectors or quality engineers to examine the supplier's quality programs and the resultant control effectiveness and product quality to ensure that the product will be satisfactory when shipped. Other large customers often exercise the same type of surveillance over their suppliers. This practice requires the suppliers to provide positive control over their own operations in a Customer Data and Support Program to avoid quality crises with their customers.

A supplier concern when customer representatives visit or reside in their facilities is how to minimize the likelihood of developing an adversary relationship with such customer personnel. Opportunities for misinterpretations and misunderstandings always exist in such a situation, even with the best of intentions on both sides and with detailed written statements of all pertinent agreements and contract interpretations. Therefore, it behooves the suppliers to act positively, where they can, to prevent any such problems from being blown out of proportion through poor treatment of the customer representatives.

Office space and facilities comparable to those of their own employees of equal or higher level rank are a minimum requirement, considering the sometimes sensitive nature of the chief customer representative's activities and records. Reserved parking may reflect a proper degree of courtesy for the customer's agent, and provision for use of conference rooms and other logistical support, where appropriate, may permit the agent to perform his functions most effectively. All of this consideration and assistance should predispose the customer representative to treat the supplier fairly in dealing with quality problems. Otherwise, what may happen if his work must be done under demeaning or irritating circumstances is that each minor problem becomes a catastrophe.

The effective support of the customer's representatives and meeting contractual provisions for timely reports, certifications, and so on typically place a requirement on suppliers for extensive records. These records must be planned for to provide the needed present and likely future information at the right time and at minimum cost to the suppliers; often they may be made part of the internally required data systems with negligible impact on costs.

Treating these special types of customers and their representatives in the positive manner they deserve, compatible with the supplier-customer interdependence philosophy expressed in 3.7 Quality Engineering and Training for Suppliers, enables the supplier to realize fully the benefits implicit in his relationship with them: fewer and unexaggerated quality problem conflicts, fewer delays in acceptance of product for payment, and resulting better cash flow and profitability.

9.10 Field Problem Predictions

Design and development organizations and manufacturing, inspection, and test groups frequently identify problems that are not subject to immediate correction. Sometimes, extensive experimentation is required to develop a satisfactory resolution of the problem; at other times, lengthy retooling is necessary or stocks of offending parts exist whose destruction cannot be justified economically. Whichever situation applies, the field will have to cope with the problem for some time.

This may require special parts, more than usual numbers of parts, special equipment or instruments, special training, more personnel, or other temporary field personnel accommodations to the problem. To support such extra effort most effectively, the field planning and operating personnel deserve as much advance warning as practicable. Therefore, as soon as such a problem is identified by engineering or the factory, the appropriate field-related personnel should be notified—with an indication of timing, quantities involved, equipment and techniques required, trends, seasonal patterns, and any other useful information they may need for their planning. As experience with the problem grows, updated requirements and explanatory information should be provided on a timely basis.

In addition to the reporting of specific problems encountered in development and manufacturing, the Field Problem Predictions program calls for the use of experimental and field data, from both in-house and external sources, to forewarn of field failures and other service requirements. The electronics industry has had available to it for some time a mass of government-developed or -accumulated failure data resulting from military and space programs. These data have been published and are available in data base form through several computer sources.

The ability to do a fair job of reliability prediction, with associated part replacement rates and other useful service measures, is thus made reasonably available to producers of electronic products. Associated mechanical and electromechanical components are typically not covered in these data bases, so their failure information must be developed by other means. Difficulties also arise in determining the effects of stress levels on the components, conditions that the designers frequently are unable to identify with sufficient accuracy.

Purely mechanical products have no such broadly available quantities of reference data. Some effort has been made to develop useful data bases, with little success to date, although a substantial amount of specialized data of this type is buried in the experience of a given producer. If these data could be collected and analyzed, that company might be able to do a satisfactory job of predicting reliability and failure patterns, even for its mechanical products.

Another device useful here is called Failure Modes and Effects Analysis, a technique that requires a thoughtful evaluation of the design to determine how each key component of the product might fail. A determination of the effect of such failure on further production, safety, product performance, or reliability helps the designer to decide where his design needs to be improved and also provides the basis for predicting failures and parts needs for the field.

A further by-product of such data analysis is guidance to factory inspection and test planners. If the quality efforts in the shop do not identify the same kinds of problems the field must deal with, the scope of such in-plant efforts should be adjusted or expanded to improve field experience. Another benefit of this program is to verify or to provide the basis for logically adjusting warranty terms.

When this program is properly used, customers can expect a minimum of stock outages, unprepared servicemen, and other irritating delays and costs involved in their receiving satisfactory service—with resulting attitudinal benefit to the producer.

9.11 Installation and Service Difficulties

When a product involves significant installation effort, as discussed under 9.2 Installation and Field Test and in all cases of service beyond item replacement, the possibility exists for the field personnel to experience problems in doing their work successfully and economically. Sometimes the design does not support diagnosis or presents physical barriers to service. In other cases components are so configured that replacement of a defective one predisposes others to fail. Both installation and service instructions are often misleading, confusing, or inaccurate with respect to the actual state of the product being worked with. And manufacturing deviations from the original design, even though permitted by the design group, sometimes cause interchangeability and other logistics problems in the field.

This list of difficulties experienced by installation and service personnel barely scratches the surface of the problem. Virtually every organization has its own horror stories to recount. When we examine the causes for the unsatisfactory situation, we find that they are basically of three kinds.

One kind derives from the organization's sometimes deliberate but always misguided failure to incorporate field inputs into the product design process. One of the techniques to accomplish this effectively is described in 6.1 Classification of Characteristics. Another is discussed in 2.9 Product Design Review.

A second kind of cause for these problems is the typically poor state of product failure information transmitted from the field. Such reports are often inaccurate, are incomplete, or reflect opinion without sufficient problem analysis. This kind of problem is addressed both in 9.12 Field Problem Controls and in 5.5 Field Quality Data Reporting Control.

The third kind of cause for field difficulties is related to the other two, but is distinct; it is the subject of this element. In addition to the above data program for product failures, we need to have a companion program to inform design and manufacturing personnel expeditiously about the types of difficulties recounted in the first paragraph of this element description. There must also be a corrective action result from such input, as well as prompt notification to the field of the action taken; both are covered by the Corrective Action Program (1.12).

To obtain this type of feedback from the field, producers have used pre-printed forms packed with the product or with the installation and service documentation. Results have been less than completely satisfactory. Others have asked for written reports, again with mixed results. One of the more effective approaches has been to abandon the attempt to obtain 100-percent response in favor of concentrating on selected installers, dealers, or customers—sometimes with direct payment or added credits to compensate them for the extra effort.

The information desired has to do with the clarity, completeness, adequacy, and utility of all instruction documents and of installation, service, and test equipment. We also want reports on the product: its received condition; its ease of installation, problem diagnosis, and repair; and its suitability for the customers' purposes. And we want to know about installation- or service-induced problems derived from the design, such as probable damage to or failure of components that must be moved or removed to permit installation or service.

Such a program of pertinent, accurate reporting from the field will enable the producer to make useful changes in design and manufacturing practice, both in the involved product and in others, and thereby to improve the speed, quality, and economy of field operations.

9.12 Field Problem Controls

When products are sold with a fixed warranty period, the producers usually have a good measure of field failure in the warranty costs they have to pay. These figures, however, have proven to be tainted in some cases by inflated claims entered by dishonest customers, installers, and service agencies. Pilferage shortages have also sometimes been disguised as warranty claims. Such possible losses point up the need for an element on Field Problem Controls as one of the six Quality System controls.

Obtaining the specific failure information has been much more difficult than getting the cost data—parts and components are replaced and discarded without analysis of failures; reports are inconclusive, incorrect, or nonexistent; service personnel are inept, in some cases damaging more than they repair. These factors, and others in specific instances, have caused many companies to despair of ever getting useful data from the field.

Even when some useful information is received by the producer despite these difficulties, it is often restricted to the warranty period. Since that period is normally significantly less than the useful life of the product, the data rarely provide any useful knowledge about post-warranty failure types and frequencies and the associated product reliability. The customer is very much interested in reliability, as it has a substantial effect on the cost of ownership of the product, so this lack of pertinent information creates a roadblock in the producer's attempt to satisfy the customer.

Many companies have come to the conclusion that they must improve their abilities in this area. Some of the techniques used are:

1. Establish captive erection, installation, and/or service facilities; train the personnel properly; establish controls on report formats and contents; and process the data efficiently. The sample of facilities selected should represent fairly the variety of field conditions, relative quantity and breadth of product, and types of applications to which the product is likely to be exposed.
2. Identify selected independent field agencies, contract with them to provide the needed information, see that their personnel are trained, standardize with them on reports, have them return the failed items for analysis at the factory, and process the data. The

independent agencies selected should satisfy the same criteria as the captive facilities do.

3. Require or promote all replacements to be made only on an exchange basis, provide for analysis of a sufficient sample of the failures, and proceed accordingly.
4. Apply an extended warranty to a representative part of the product line, thus getting reliability data that may be used for much more of the overall product line.

To combat the improper use of warranty, producers have run statistical studies of claim patterns from field agencies and large customers, instituting appropriate managerial or legal action when cheating is detected. They have also run prepared samples through suspected agencies to detect both error and fraud. Error leads to retraining, equipment calibration or replacement, or clarification of instruction; fraud is appropriately handled.

A related problem occurs when the service agency does not correct the customer's problem the first time or causes a second problem while correcting the first. The advantage in this case lies with those companies that have durable products unique to a customer or serially numbered and that can identify the service information to such individual piece of equipment. One device often used to detect this situation is to set a time limit, often thirty days, within which a second service request will be charged to a lack of correction or to damage from the first. Of course, if the customer complaint is recorded, a repeat complaint would be viewed in the same fashion.

A number of successive service calls for the same reason would indicate service ineptness or, if widespread, a probable design failure to support diagnosis. Frequent service requests on the same equipment might reflect lack of service capability, but might also indicate a design failure to consider service requirements, equipment, and practices. Analysis of the repetitive service calls should indicate the proper areas for correction.

By whatever means field data of a satisfactory confidence level are collected, their further analysis can provide several types of useful results:

The basis for useful revision of warranty terms or duration, including design trade-offs

The occurrence frequency ranking of workmanship nonconformity classes for retraining and control purposes

The basis for selective design revision for desirable reliability and serviceability improvement

The validation of reliability testing in the laboratory, particularly accelerated life testing

When the proper measurements are made of field failures and service patterns, many of the areas of customer dissatisfaction become visible. They can then be attacked with the proper permanently remedial programs, thus greatly improving the producer's long-term acceptability in the marketplace.

9.13 Field Problem Handling

The problems of obtaining useful failure and other data from the field are explored in the program of 9.12 Field Problem Controls. Often, however, these data must be supplemented with a program of intensive analysis of the failed items to determine the real sources of the failures. Whether this analysis is performed by technicians in the field or by personnel at the factory is determined by applying business, relative capability, diagnostic facility, and other factors to the decision. Whatever approach is employed, the Field Problem Handling process must be rapid and accurate in all its phases, from failed item return to action-oriented report of findings.

Failure analysis is often expensive, with little additional knowledge obtained from processing high volumes of items. It is therefore common to conduct analyses only on a sample of failures. Techniques used to obtain useful valid samples include analyzing only those items processed by selected service agencies, having only those items bearing certain last digits of their serial numbers analyzed, or analyzing only those items exhibiting special failure modes or coming from certain customers.

Another source of information useful in identifying problems is examination of the field reports generated in 9.12 Field Problem Controls. Detectable trends may indicate small but growing problems that can be attacked and corrected before they reach catastrophic proportions. In the special case of custom-built products, very few if any of which are alike, evaluation of combined failure information may be practicable only when confined to common parts or modules. Such analysis can often be very useful in revealing unexpected patterns of failure.

When, having determined through analysis of failed items or of field data that there is a significant quality, reliability, or safety problem, the producer must proceed with the appropriate response. The response may be entirely within the producer's power to determine, both as to extent and as to timing, or it may be mandated by external regulatory agencies. Regardless of the circumstances, the company should have its response program prepared in advance, with the specifics developed as soon as the problem is identified.

Typical responses include product recall and repair, exchange, modification on site, or provision of "do-it-yourself" kits to customers. The response mechanism should also include revision of design drawings and specifications or manufacturing practices, as appropriate. Where design responsibility is involved, there should be a provision to prevent future product designs from producing a similar problem.

Regardless of the source of the problem information or the company's relative freedom to develop its responses, the response must be rapid. This includes determining the cause of the problem, its effects on customers, the extent of customer involvement (numbers and identification), the development of the corrective actions to be taken and by whom, and the communication of the fact of the problem and the details of its resolution to customers, dealers, and field service personnel. Any tooling, measurement equipment, or special parts—with associated instructions and training—must also be provided rapidly, consistent with assurance that the work will be properly performed. Often this can best and most expeditiously be handled through widely available computers by which the field personnel can rapidly transmit their data to the control source and receive advice, further data (perhaps from other locations), and action authorization when needed.

This program's concern for field problems is not limited to product failures and safety hazards, although, because of customer involvement, these are the most important areas. The program also includes customer use difficulties that are correctable by the producer's efforts.

In this category falls the handling of customer complaints. Since complaints may be received by dealers, distributors, service personnel, marketing, sales, and, in fact, practically anyone in the company, there must be a well-developed and disseminated plan for dealing with them. The plan must be one of the bases for training everyone potentially involved and for the audit program of 9.15 Audit of Field Quality Activities. The plan must also provide easy access for the customer, a central accumulation and analysis point, rapid resolution devices, and active solicitation of customer feedback of handling effectiveness.

Constructive thought should be given to presenting any public exposure of problems and their solutions positively. Many organizations fail to recognize, and thus to publicize, the fact that their actions in correcting problems, particularly when the product is out of warranty, are evidence of good citizenship in the marketplace. These corrective actions sometimes extend the life of the product or advance the state of the art and consequently are valid sales devices, if properly presented to the market. All documents and public releases associated with any such problems and organizational responses to them should therefore be prepared and reviewed with an eye to capitalizing on the apparent adversity.

Having applied a carefully developed program for handling field problems, a company will realize reduced frequency, severity, and cost of field problems and may even be able to turn them into assets.

9.14 Customer Satisfaction Measurement

Measuring and improving customer satisfaction with the product or service is an important consideration for any organization that wishes to remain in business. To do so requires that the producer determine specifically just what needs to be measured to obtain a valid indication of customer satisfaction. Often this can only be accomplished by getting representatives of "the customer" involved in helping to develop the required measures.

Although difficult to obtain accurately, Customer Satisfaction Measurement is another of the six basic controls of the Quality System. One factor contributing to the failure of many companies to measure it successfully is that the very method chosen to elicit reaction information may change the customer's response. Personal interviews, potentially the most effective information-gathering technique, are most likely to alter the customer's stated attitude. The personality and capability of the interviewer, the fact that the interview generates a "They really do care!" response, any third party or audience interaction, common unwillingness to admit other than impersonally that "I bought a lemon!"—all of these influence response to an interview. The design and control of the interview process and the selection and training of interviewers must be very carefully accomplished to ensure success of the program.

Other companies depend on questionnaires delivered with the product or mailed immediately after registration or at some fixed time thereafter. The mailings may be to a sample of customers or to all. In any case, the returns are highly variable in number and utility, both dependent on the wording, length, and timing of the questionnaire and both affected by the cover letter or other explanatory material.

A third source of useful information may be customer complaints or other unsolicited customer input. Combining this material with the solicited reaction data, however, can be very difficult.

The fourth primary source of these data is the field service organization, whether company-owned or independent. Much of their data is associated with inputs from Field Problem Controls (9.12) and is subject to the same difficulties of volume, accuracy, and utility suffered by the Controls programs. Programs using selectively generated sample data, however, frequently enable companies to obtain sufficient valid information in both these areas to permit making proper decisions—particularly when the data are appropriately broken down by customer type so that specific and relevant improvement responses can be developed.

Procedures exist for combining information from two or more of the above types of data gathering into a single numerical measure of customer satisfaction, often with weighting factors reflecting relative validity of the source data. This permits trends to be identified (when the frequency of measurement is supportive of trend determination), competitive and long-range proprietary comparisons to be made, and major problem areas to be highlighted for corrective action.

Careful planning and validation of the measurement program are essential because of the substantial potential for measurement error, the ultimately vital nature of this information to the future success of the enterprise, and the considerable cost involved in some of the techniques, particularly extensive interviews. Importance to success implies, of course, that corrective and improvement programs will be initiated when product, marketing, or service problems or opportunities are identified.

While the above discussion was oriented to measuring customer satisfaction, it is important to keep in mind some of the principal causes of customer dissatisfaction. These include delays in receiving ordered items or services; receiving wrong items, quantity, information, or support; being faced with incivility or incompetence by the service provider; or being asked to pay more than originally agreed upon or to lose the use of the item or service through no fault of his own. When Customer Satisfaction Measurement is attempted, it can be useful to try to capture an indication of which customer irritants are operating in our situation.

Having a validated ongoing measure of relative customer satisfaction with the company and its products and services, management can guide product design, production, service, marketing, advertising, and public-relations activities to achieve maximum marketplace acceptance of its efforts. A further benefit is improved long-term customer loyalty for the product, brand name, or service. As a result, the company can expect to be recognized by its customers, trade associations, and others for its accomplishments in improving the quality of its products and services.

9.15 Audit of Field Quality Activities

Manufacturers and service organizations take many different approaches toward answering the questions associated with parts and finished goods warehousing, distribution, installation, and service. The control of these items and activities also takes many forms. The primary intent of such control programs, however, should be to minimize cost of these activities and maximize customer satisfaction. The producer, even though dealing with the ultimate user through OEM customers, through independent distributors, installers, or service agencies, or through franchisers, also has a legitimate concern in protecting his reputation because the user may identify him with some unsatisfactory aspect of the product or service, despite the producer's actual noninvolvement. Therefore, it is necessary to have some type of controlled, objective audit of the quality of finished goods, service parts, installation and service performance, and relevant software. The audits may have to be conducted by outside agencies, by intercompany teams, or by producer personnel, but they must be governed by sound audit principles, as developed under 1.14 Audit of Procedures, Processes, and Product.

The Audit of Field Quality Activities plan should cover all hardware anywhere in the chain from the producer to the ultimate user, except items awaiting or in process of internal production. This includes spare and replacement parts from suppliers going directly into the service channel; internally produced parts, subassemblies, and assemblies doing the same thing; finished product, including software, released from production; and packing and packaging material, its usage, and its disposition. In addition, the audit plan should include the comparison of activities—performed by material handlers, warehouse workers, truckers, installers, service personnel, and others who can affect product and service cost and quality—with those specified in or by appropriate manuals, bulletins, performance and quality standards, and other directives. One plan should cover all audits, including those that may result from the programs of 4.5 Packaging and Packing Control and 9.7 Drop-Shipped Item Control.

The audit plan will, of course, identify who is to perform the audit, and how; but it should also contain clear instructions concerning investigation of nonconformances, reporting of results, and the obtaining of indicated permanent correction of errors in requirements and of the root causes for nonconformance.

As with all recurrent audits, the plan should contain an objective scoring system so that level and trend comparisons can be made among agencies, product lines, personnel, or any other significant groupings. It must also expedite proper warning about and correction of any unsafe conditions and practices, impending product and service quality catastrophes, and potential or real negative environmental effects identified through the audit or otherwise encountered.

Audits of the quality of complete product, spare parts, and service items and audits of the actions of sales personnel and of the service provided against standards and procedures give management the necessary information to take action to ensure that the customer develops and maintains a positive attitude toward the company.

Section 10

Quality System Engineering

In developing and applying the Quality System for an enterprise, an overall end date should be established. Then the following steps should be taken, roughly in this sequence:

1. Determine the extent to which the elements of the nine subsystems will benefit the business; some probably will not apply at all.
2. Identify and describe in writing any activities that should be part of *your* Quality System but are not covered by any of the elements in the subsystems; there may be one or two such activities, although typically they can be covered by logical extension of the scope of the listed elements.
3. Evaluate the extent to which present activities satisfy all the Quality System requirements for each of the active elements (those resulting from Steps 1 and 2).
4. For the active elements not fully covered by current documented programs, establish a rank order of priority and schedule for development or expansion.
5. Develop written procedures, training programs, technical documents, or whatever else might be needed to satisfy the organization's needs for documentation of each of the active elements.
6. As each item is developed, train involved personnel and test it in a suitable environment to assess its effectiveness and economy and alter it as indicated.
7. Apply each program broadly as it is proven effective and according to a realistic schedule, including necessary training for all initially and ongoing, as required.

Steps 1 through 4 result in a fairly rapid configuring of the needed Quality System for the enterprise; the remaining steps are carried out for each element individually and take varying amounts of time from one company to the next dependent on the number of elements, their degree of current coverage, and the resources made available for the work. The time needed for Step 7, of course, is influenced by the organization's resistance to change and its ability to absorb changes. Each of the seven steps can be aided by a structured approach, suggestions for which follow.

Steps 1–4 (Evaluation)

Steps 1 through 4 may be done by a knowledgeable individual or, more effectively for many reasons, by a multifunctional group of knowledgeable individuals performing either each step separately or all four at once. To save time and to provide a broader base of involvement and understanding, combining the four steps and using the multifunctional approach is superior to the alternatives. The following discussion assumes that this is the chosen method.

Work sheets should be prepared to support the establishing of priorities in Steps 1 through 4. Organizing these by subsystem could be useful, since some of the functions and individuals involved in the process of setting priorities could properly vary by subject. In addition, this approach will be helpful later on in Quality System Management.

An example of such a work sheet, providing space to tabulate all the information needed for the establishment of priorities, is shown in Exhibit 1. There are two parts to the rating process, labeled "System Need Satisfaction Level (N)" and "Degree of Visible Benefit (B)." The two parts should be rated in sequence for each element of the subsystem. For the "Need" rating, the ingredients are:

- Should we do anything in this area?
- If so, are we doing all of what is called for now?
- If so, how well is it documented?

If the answers are "yes," "yes," and "very well," the rating given would be Outstanding and shown in the 0 column. If the answer to the first question is "no," the element would not be rated and a note would be made in the "Comments" column. Any other combination of answers would receive a higher number resulting from the application of the following definitions (or something similar).

The six "Needs" rating categories other than "None" might be defined by an organization as:

- Outstanding—Applying all of the principles contained in the element description and documented for use throughout all pertinent portions of the organization, with excellent and sustained results everywhere
- Qualified—Comparable application throughout the key portions of the organization, with demonstrated good-to-excellent results in those areas
- Marginal—Comparable application in many of the key portions of the organization, with numerous positive trends resulting from that application
- Fair—Applying some of the principles, not well documented, in a few of the key portions of the organization, with some resultant positive trends
- Weak—Little application of the principles, not well documented, with few results observable
- Poor—Little application of the undocumented principles with nothing of moment accomplished

The second, or "Benefit," rating would be determined only if the "Need" rating was other than 0, and it would depend on the following questions:

- If we did (or are doing) what is called for, how big would be (or are) the results?
- What would be (or is) the effect seen by the customer or our own operations?
- What is the measurable impact on our Quality System?
- How soon will we see the results from doing all that this element prescribes?

In this case, a 0 rating would be awarded only after determining that nothing would result from doing whatever of the element is currently not being done. A rating of 1 indicates that the effect of the element would be beneficial, but it eludes measurement—it is "Intangible." Any other rating is the group judgment of what and how visible the results would be, considering all the relevant questions.

EXHIBIT 1 Element Priority Work Sheet

ORGANIZATION: DATE: SUBSYSTEM:			*System Need Satisfaction Level (N)*							*Degree of Visible Benefit (B)*						SCORE	PRIORITY
Element			None	Poor	Weak	Fair	Marginal	Satisfactory	Outstanding	None	Intangible	Slight	Moderate	Substantial	Very Large		
No.	*Title*	*Comments*	10	8	6	4	2	1	0	0	1	2	3	4	5	NxB	

When measurements have been identified in the process of rating each element, they should be shown in the "Comments" column for future reference.

The number scales shown have proven useful in practice but are not mandatory; intermediate values can be used in the "Needs" section of the form to reflect appropriate modifications of the ratings—based on the comments for a given element or on other judgment criteria. The numbers selected should be placed in the proper columns for each element.

Priorities are assigned based on the scores for all elements, with the highest scores (maximum is 50) receiving first attention. Elements with scores of 5 or less normally would not be included in the list for any work. Those scoring below 20 would usually not be given early attention, unless they fall in the category covered in the next paragraph. If it is necessary to break ties because of limited resources, the rating group should select the element(s) most likely to provide immediate visible benefit to the enterprise.

A problem arises in the case of those elements that are frequently given an "Intangible" rating for Category B, Benefits. Every effort should be made to avoid such an assignment, but some elements are inherently incapable of benefit measurement. The few that always achieve this rating are usually vital to the successful operation of the Quality System and require early attention for that reason. After the priority process has proceeded through Steps 1, 2, and 3, it is necessary to examine the Intangibles carefully and to increase their priorities arbitrarily, if this appears necessary. Obviously, if their "Need" score is 0 or 1, there would be no reason to adjust their priorities. But if the Need is high, such a move could well be required. Guidance in this case can be obtained by referring to the asterisked questions in the Subsystem Evaluation Work Sheets of section 11 (page 292). The corresponding elements should be upgraded in priority according to the foregoing criteria. Additional discussion of the subject of "Intangible" elements may be found in 1.15 Quality Program Effectiveness Review.

To be successful, every system requires specification, measurement, comparison with standard, feedback of results, and corrective action where required. In the Quality System, specification is provided by procedures, objectives, and standards—with many other specifics, such as drawings and customer requirements, included by reference. Measurement, comparison, and feedback of results are accomplished by the six control elements of the System: 1.7 Quality Cost Program, 1.14 Audit of Procedures, Processes, and Product, 2.10 Customer-centered Quality Audit, 4.13 Yield Control Program, 9.12 Field Problem Controls, and 9.14 Customer Satisfaction Measurement. These elements are supplemented by 1.15 Quality Program Effectiveness Review, 5.3 Received and Produced Quality Data Reporting, 5.4 Appraisal Activity Evaluation Data, 5.5 Field Quality Data Reporting Control, 7.7 Audit of Quality Measurement and Control Equipment, 8.8 Training Effectiveness Measurement, 8.12 Quality System Personnel Performance, and 9.15 Audit of Field Quality Activities. Corrective action is initiated through all these elements, plus many others in the System, which all culminate, when initially unsuccessful, in 1.12 Corrective Action Program.

Every Quality System tailor-made for a manufacturing enterprise needs the six controls, or something like them, and most or all of the other elements identified in the preceding paragraph. Purely service organizations may not need some of them; therefore, the list of "must be included" Intangible elements cannot be completely stated here. The only criterion is that the five requirements listed at the start of the preceding paragraph must be adequately satisfied.

In many cases, the number of elements truly deserving priority attention is quite large—more than 25 or 30. In such situations (and even in some with fewer priority elements than 25), it is often desirable to consider combining related elements into "project" packages. This reduces the administrative workload to a degree and makes for easier description of the areas covered. However, exercise caution: The scope of the "project" may be too great, making it difficult to define all aspects of the relationships among the elements. Recognize, though, that people not intimately involved in the development effort find it easier to follow the progress of 15 "projects" than 50 elements.

Step 5 (Development)

The writing of the documents associated with the priority elements may be accomplished in many different ways, each having advantages in certain circumstances. Some organizations assign capable people part-time from the involved organizational components: Quality, Engineering, Manufacturing, Purchasing, Personnel, Service, Finance, Marketing, and so on. Others assign full-time people representing the same functions. Combinations of these two approaches are seen. The work may also be done by contract with outside agencies, by the internal Methods and Procedures group, or by college students or others hired just for the project. The method chosen should, whenever possible, be selected based on its training potential for the writers and its effect on the ultimate utility of the resulting documents, including how readily they will be accepted by the entire organization. Each element or project may be assigned to an individual or to a "writing group." The leader of the group would also probably be responsible for guiding or directing the implementation of the resulting document.

Another aspect of this question is how the writing is done. Input on each subject is required from each of the functions involved, as to both what current practice is and what the new practice should be. Whenever practicable, current good practices should be used—at least as the basis for the new program—minimizing disruption of and rejection by the organization as the new program is introduced. As much as practicable, the new practices should be consistent with the "culture" of the organization by which they will be used. Documents prepared at this stage should be clear, complete, auditable, and concise, should satisfy the Quality System requirements, and should fit the format selected.

Procedures should be in Playscript (see Appendix A), and technical bulletins, or whatever they may be called, should be properly issued, along with any other type of pertinent information. Wherever possible, these documents should contain a list of the mechanisms by which the effectiveness of the new practice is to be measured. Such mechanisms should be based as much as possible on existing management or operating measurements.

Some organizations reject formal written procedures for a variety of reasons. In these circumstances, documentation might be limited to a listing of key points for control purposes. Everyone could then exercise his own ingenuity as to what was done and how, being required only to satisfy the key point criteria. This approach is compatible with the entrepreneurial and job-enrichment approaches to personnel management. Preparation time for this type of document is not much less than that for a complete procedure, since at least the flow chart must be generated to ensure that the writer has not overlooked a key point.

Another type of document that has a major role in the Quality System is the technical bulletin or memorandum, which normally does not impose auditable requirements in the way a procedure does; instead, it provides extensive technical direction to the organization. Every element of Subsystem 6, Special Studies, could be the subject of one or more such bulletins, with no procedures written for them. Other elements could also be handled this way. When it is desirable to make any of these or other pertinent documents mandatory, they may be incorporated by reference in an appropriate procedure.

The contents of training programs, on the other hand, will be little different regardless of whether the organization uses procedures, but they will have to be provided by experts. If a sufficient level of expertise does not exist internally at the start of the program, it will have to be provided somehow, perhaps through outside agencies or through in-house study.

Step 6 (Implementation)

The testing process is important for any subjects completely new to the organization. Even major changes to existing programs may have hidden flaws or unexpected violations of local "culture" that can be very expensive to overcome or can make the program ineffective. It is therefore simply good business to conduct pilot runs on these items to be sure of their utility and economy. This must be done with the prior agreement of the involved groups and with built-in detection and notification methods to cover any problems. The document writer or writing group leader should be personally involved in these tests or, at least, should be in frequent contact with the personnel taking part in them.

In developing the written plan for the Implementation Phase, it is important to select areas or projects that will thoroughly exercise the new or modified practice to the fullest possible extent. To ensure that this is accomplished, the plan should identify the individual(s) who will lead this phase. It should include a schedule for frequent assessment of progress toward complete application of the new practice in the pilot-run situation, as well as a list of actions to be completed according to that schedule.

The list might include the area or activity to be used for the tryout, development, and conduct of training programs for the people in the tryout project, as well as dates for starting and completing each part of the procedure. Any portions of the documents that prove less than satisfactory should be corrected and a plan developed for the Installation Phase. If possible, the measurement mechanism(s) developed or identified in Step 5 should be used to provide an initial look at the effects of the new practice so that an estimate can be made as to what the fully installed effect will be.

The testing activity results either in alteration of the program when indicated or in its release for full-scale application.

Step 7 (Installation)

In some circumstances it is appropriate merely to release the document for general use. In others, training programs, perhaps covering several documents in one program, must be conducted before such release. In still other cases, the extent of the undertaking is so large or available resources are so limited that it is impracticable to attempt to apply the program everywhere at once. An example of this might be a program to be applied to all suppliers, of which the enterprise might have 2,500; even if two suppliers could be involved each day, it would take several years to cover them all—not even considering any new ones acquired in the meantime. Involving groups of suppliers in meetings might reduce the total time; but they would still have to be listed by priority, scheduled, and worked with individually to resolve problems, so the total time would still be substantial.

A trade-off, then, needs to be made between the application schedules and the realization of the benefits of this type of program. This situation should be identified early in Steps 5 and 6 and plans made and published before beginning Step 7 for each of this kind of element. These plans should be compatible with the overall Quality System development and application schedule established initially; and resources will have to be made available to satisfy these time constraints, assuming they are valid.

Whatever the size of the organization, significant personnel resources (which, in this assignment, would be charged as prevention costs) and elapsed time are involved in any program to build a Quality System. It is hoped this section will have contributed to the success of that effort. In the next section, on Managing the Quality System, an attempt will be made to explain what to do with the Quality System after it has been built.

Section 11

Managing the Quality System

Since all manufacturing and many service organizations have Quality Systems of some sort, and since the development of a Quality System of the extent described in the preceding chapters produces some of its documentation relatively early in the Development process, management of the system is of concern at all stages—not just after the Installation effort is complete for all elements. In addition, in large companies there are basically two levels to this management: the local level and the remote (sometimes corporate) level.

The management of the Quality System is the responsibility of the Chief Executive Officer (CEO) of the enterprise. While the CEO may delegate aspects of the effort to others and, in the interest of building a sense of ownership for pieces of the System among all members of the organization, may measure their performance against approved accomplishment goals, the ultimate responsibility is still his. Recognizing this, the CEO should appoint a Quality System Council or Steering Committee, composed of key senior managers from each major unit, who will be responsible to him for seeing that the System is managed effectively within their operations. The committee will take an active role in planning for and seeing that resources are allocated to quality improvement, conducting and receiving training in the Quality Sciences, engaging in 6.6 Problem Solving, and ensuring that their direct and indirect subordinates do so as well. It will also communicate with employees, customers, suppliers, and others on quality subjects.

At a local level, then, the top person either will be a member of or report to a member of the council or committee, probably using his leading

Quality function manager as his executor. The remote level manager will probably be that person reporting to the CEO who is a member of the council or committee and has a full-time Quality assignment. For the remainder of this discussion the remote level manager will be called the corporate Quality System manager.

The CEO is responsible for seeing that the operation of the Quality System receives the same strategic treatment as other key business activities. This can often be assisted by incorporating Quality System action requirements into the measurement and awards program for the organization. When the Quality System objectives are given equal or greater emphasis than that given shipments, profits, and other key measurements, the message will be clear that managing the System is of high priority.

At the local level, managers of all functions are concerned with the day-to-day applications of the Quality System provisions in their activities. This requires everyone's thorough familiarity with the relevant portions of all procedures; the availability to all personnel of the pertinent procedures and other related documents; the assignment of sufficient capable personnel to perform the work; and the regular generation, analysis, and timely reporting to all of those data that reflect the current condition of each control measurement. In addition, as improvements are achieved, steps should be taken to ensure that they become permanent and then serve as the basis for further improvement.

In carrying out these responsibilities, each functional manager must develop plans to support the unit's Quality System priority objectives. The planning process in many cases can benefit from contributions by other managers and employees, suppliers, dealers, and even customers—both internal and external. The plans should contain statements of the objectives, the sequence of actions to be taken, who is to perform each action, what other resources need to be applied, the schedules with milestones, the mechanisms for measuring success (including the standards or benchmarks to be used for comparison), and the relationships with other improvement plans within the unit and any larger organization.

The managers who produce the plans must ensure that the resources—both personnel and other—are provided. This may require that capital expenditures be requested and approved; that parallel plans be made for hiring, training, and/or assigning people to the project, with associated methods for continuing improvement of their effectiveness and productivity; that employee feedback from and to the process be obtained, perhaps through team practices, with ready access to upper levels of management; and that the participants be encouraged to achieve innovation.

Job descriptions should contain the appropriate key points of the Quality System responsibilities, and the incumbents' performance should be measured against the pertinent ones for personnel development and any management incentive purposes. As the System is developed and used, the job descriptions will have to be modified to keep pace. This also applies to the procedures, themselves, for which coverage should be included in 1.5 Documentation Control.

When either internal or external customers are affected by any of the projects of the unit, it is essential to follow up with them to ensure that all aspects of the service are and continue to be satisfactory. If not, the situation must be analyzed and corrective action taken. This is especially true if the action was the result of a customer complaint in the first place. Any corrective action taken in the case where the complaint is not fully resolved must be addressed and the complaint handling process improved to avoid future problems.

Since much of the success of the Quality System depends on the effective training of the personnel involved, a permanent training effort as covered in Subsystem 8 will be required. Newly hired and reassigned personnel will have to receive appropriate training, and repetitive refresher training will be required for many experienced employees. The enterprise must recognize this need and make the corresponding investment in this key prevention activity.

Audits of all types must be scheduled, held, and the results followed up. Since many of the audits will involve multifunctional teams, provision for obtaining and rotating the audit team personnel must be made.

The local Quality System manager is responsible for overseeing the work of the System in all areas and reporting its accomplishments and shortcomings to the local council member.

The corporate Quality System manager is frequently concerned with encouraging and supporting the introduction of the Quality System in a number of operating units at the same time. To do this effectively requires having the tools to evaluate the health of the Quality System in each unit, as well as the tools for measuring changes and accomplishments. For that purpose, the Element Priority Work Sheet (Exhibit 1 in section 10, Quality System Engineering) can be applied for every unit. The corporate Quality System manager will find the work sheet helpful for obtaining a picture of the overall strengths and weaknesses of the enterprise. On the work sheet he would list those areas where his staff might be able to contribute technical bulletins or other central publications, thus eliminating the need for them to be developed several times.

Another approach for evaluating the continuing health of the Quality System, similar to that of the Element Priority Work Sheet, would be to look at the condition of each subsystem and at the complete System. The work sheets are shown as Exhibits 2 through 11 in this section. The concepts on which both of these approaches are based were developed by Allis-Chalmers in its renowned Quality Assurance program and were expressed to me by William Kohl, then corporate director of that company's Quality Assurance Service. The questions on these exhibits may need to be augmented, replaced, or modified as time passes and circumstances change. In all likelihood, many of them will have to be reworded to accommodate their use in service industries, government applications, and institutions. As much as practicable, the intent and concepts expressed in the original questions should be retained. Also, it is important that, irrespective of the change, the most important half of the questions carry asterisks.

In practice, the corporate effort would involve audit and support for the local Quality System activities. The audit would be performed as described in the "procedure" portion of 1.14 Audit of Procedures, Processes, and Product but would use Exhibits 2–10 as the audit checklist. Specific criteria must be established within each company to determine what constitutes achievement in each rating category. Since the descriptive terms proposed below are modifications of those used in rating the "Need" factor in the Evaluation stage of Quality System Engineering (see page 278), they should provide a good basis for comparing results of the System Installation with the conditions existing at the time the initial evaluation was made.

The six rating categories might then be defined by the organization as follows:

- Outstanding—Applying all of the principles of the Quality System element(s) involved throughout all pertinent portions of the organization, with excellent and sustained results everywhere
- Qualified—Applying those principles throughout the key portions of the organization, with demonstrated good-to-excellent results in those areas
- Marginal—Applying those principles in many of the key portions of the organization, with numerous positive trends resulting from that application
- Fair—Applying some of the principles in a few of the key portions of the organization, with some resultant positive trends
- Weak—Little application of the principles, with few results observable

- Poor—Nothing of moment accomplished in the area of the involved Quality System elements

(*Note:* This list, while derived from actual experience, has been adjusted to be somewhat compatible with the 1989 Scoring Guidelines of the Malcolm Baldrige National Quality Award.)

The corporate audit could serve as a preliminary evaluation, to be followed by an in-depth investigation by the local group—reflecting a two-stage approach, with the second stage thus less demanding of resources than if the two efforts were completely duplicated. Before conducting the audit, the auditor(s) would fill in the Applicability columns (each of which must total 100) and would enter the factor ratings as each evaluation step is completed, using the above criteria. Sufficient time must be allowed for the audit to permit thorough evaluation of all factors; otherwise, the results will be highly questionable.

In filling out the work sheets, numbers other than those shown for the rating categories would not be used. While the Applicability totals must all equal 100, the values for each question would probably differ among units of a large organization, perhaps by as much as or even more than among companies in one industry. The results of the evaluation of all the subsystems would then be posted to Exhibit 11. Exhibit 11 contains a provision for a Relative Importance factor by which the Subsystem Ratings are multiplied. The sum of these multipliers must be 9.0, equaling the number of subsystems. In most cases, all nine values would be 1.0, but there may be special circumstances, particularly for service industry applications, where variation in these factor values would be proper.

After all the tabulations are completed, the overall System rating is calculated. Comparisons with previous ratings and with those for comparable units would be made and guidance and assistance provided for improvement. Annual improvements in rating would be expected, with the intent of reaching "Qualified" status according to an agreed-upon schedule. The questions identified with asterisks on each work sheet are considered critical to the success of a Quality System and must *all* be rated as "Qualified" or "Outstanding" for the unit to reach full qualification.

In the Allis-Chalmers program, achievement of a Qualified status from the corporate audit, accompanied by a substantial reduction in quality costs and by improved field quality experience, resulted in an award of the cherished Total Quality Assurance Certification. Reaching this objective on schedule had a significant effect on management bonuses; failure to reach it could have had a serious effect on management job security.

A comparable commitment by any organization will result in obtaining the benefits of the Quality System: reduced costs and increased customer satisfaction and acceptance. The local Quality System manager can use the same audit approach as corporate, including the work sheets, as his own method for highlighting areas needing improvement.

The support aspect of the corporate program would take the form of technical assistance to the local Quality System development task force, provision of guidelines for local use (perhaps similar to Appendix A), preparation of technical bulletins for the entire company, distribution of explanatory material about the Quality System and how to install it, and personal assistance to the local Quality System manager in overcoming any obstacles in the introduction and management of the System. This last could include taking part in local meetings, helping recruit specialized personnel, and so on.

One of the problems with any good concept is that, after it is successfully applied, people tend to forget what things were like before it was introduced. Many times we have seen management concern for short-term profitability result in the elimination of critical longer-term or already successful activities. Indeed, why should one continue the Prevention investment when Failure costs have virtually disappeared? The corporate Quality System manager must face this question squarely with the Council, just as his local counterpart must deal with the Council member to whom he reports.

Often it is proper to reduce, otherwise alter, or even eliminate appraisal efforts when either the permanent corrective action or new product- or process-related prevention efforts have proven effective (see 6.6 Quality Level Improvement). Occasionally, circumstances may change so that some prevention actions are no longer useful. Indicated changes should be made, but not under the threat of mandated expense reduction, which almost always results in more total cost—particularly when the prevention activities that were cut were effective.

Both the corporate and local Quality System managers have the responsibility, then, of ensuring the ongoing validity of their programs. (See 1.15 Quality Program Effectiveness Review.) Even this will not suffice in all cases to fend off the threat of blanket expense reduction. Therefore, it is essential from time to time to publish reports reminding all managers of the specific cause-and-effect accomplishments of the System to date. Such reports should take some form that is dramatic and highly informative, leaving no doubt of the System's utility. Even though such reports may correctly be viewed as self-serving, they should still be issued if the contents cannot be successfully contradicted. Since the Quality System involves every function in an organization, its success reflects favorably on all of them, not just the Quality one. Thus, the Quality System managers are responsible for ensuring the continuation of effective System effort by all appropriate means, including politically based ones.

When the whole Quality System is operational, virtually every element will support all the others, making each of them more effective than it would be alone. The net effect of this is a quantum leap upward in customer satisfaction and economy, which is only partly realized as each element is put in place. Continuing positive attention to all aspects of management of the Quality System will then produce unanticipated dividends for the organization.

EXHIBIT 2 Evaluation Work Sheet: Quality System Management

ORGANIZATION: DATE: SUBSYSTEM: Quality System Management		*Factor Rating (R)*						APPLICABILITY	SCORE
	Factor	Poor	Weak	Fair	Marginal	Qualified	Outstanding		
No.	*Description*	10	8	6	4	2	0	(A)	RxA
* 1	Does the Quality Manual contain current policy and procedures?								
* 2	Are System Audits conducted and followed up?								
* 3	Is the Corrective Action Program effective?								
4	Is material traceable as required?								
* 5	Are Quality Costs used as a primary tool for resource allocation?								
6	Are Quality Objectives set and measured against for all functions?								
* 7	Is application of the Quality System on schedule?								
8	Are the current programs of the Quality System effective?								
9	Are customer quality program requirements satisfied?								
10	Is the Quality organization capable of fulfilling its Quality System responsibilities?								
								100	
							Subsystem Rating		

* See page 280.

EXHIBIT 3 Evaluation Work Sheet: Product Development Control

ORGANIZATION:

DATE:

SUBSYSTEM: Product Development Control

No.	Factor Description	Poor 10	Weak 8	Fair 6	Marginal 4	Qualified 2	Outstanding 0	APPLICABILITY (A)	SCORE RxA
*1	Are complete schedules set for all new product introduction activities?								
*2	Are products researched to be responsive to customer requirements?								
*3	Are product design reviews held regularly and effectively?								
4	Are all product designs fully producible when released, with established satisfactory yields and demonstrated reliability and safety?								
*5	Are customer-centered quality audits regularly performed and the results acted upon?								
6	Are all product approvals received from customers and others on schedule?								
7	Are test methods fully explored and determined to be adequate?								
8	Are the make-or-buy decisions made with full consideration of quality implications?								
								100	
	Subsystem Rating								

Factor Rating (R): Poor = 10, Weak = 8, Fair = 6, Marginal = 4, Qualified = 2, Outstanding = 0

* See page 280.

EXHIBIT 4 Evaluation Work Sheet: Purchased Material Control

ORGANIZATION:

DATE:

SUBSYSTEM: Purchased Material Control

Factor No.	Factor Description	Poor 10	Weak 8	Fair 6	Marginal 4	Qualified 2	Outstanding 0	APPLICABILITY (A)	SCORE RxA
1	Do suppliers receive quality information packages as scheduled, with replacement of obsolete items?								
2	Are purchase orders placed only with qualified suppliers?								
* 3	Is the Supplier Rating Program used fully to support supplier choices?								
4	Is Supplier Certification used with at least 20% of suppliers?								
5	Do all lots receive complete disposition within three working days after receipt?								
* 6	Are corrective action requests properly issued to and answered by the suppliers?								
* 7	Is Quality Engineering support provided to suppliers when required?								
* 8	Have purchased material quality levels improved adequately over the previous year?								
								100	
	Subsystem Rating								

Factor Rating (R): Poor 10, Weak 8, Fair 6, Marginal 4, Qualified 2, Outstanding 0

* See page 280.

EXHIBIT 5 Evaluation Work Sheet: Process Development and Operation Control

ORGANIZATION: DATE: SUBSYSTEM: Process Development and Operation Control		Factor Rating (R)						APPLICABILITY	SCORE
	Factor	Poor	Weak	Fair	Marginal	Qualified	Outstanding		
No.	Description	10	8	6	4	2	0	(A)	RxA
1	Are process design reviews conducted as scheduled and the results acted upon?								
2	Are all processes and appraisals properly specified and the documents fully distributed?								
* 3	Are all special processes adequately controlled?								
* 4	Are all appropriate quality checks built into the manufacturing processes and equipment?								
* 5	Are complete manufacturing quality plans developed and applied as planned for all products?								
* 6	Are all violations of forecast yields successfully subjected to the yield control program?								
7	Are factory housekeeping and safety satisfactory?								
8	Are all manufacturing processes in statistical control at satisfactory levels?								
9	Is packing and shipping fully oriented to ensuring customer satisfaction with the received product?								
* 10	Are all specified product audits conducted?								
								100	
								Subsystem Rating	

* See page 280.

EXHIBIT 6 Evaluation Work Sheet: Quality Data Programs

ORGANIZATION: DATE: SUBSYSTEM: Quality Data Programs		*Factor Rating (R)*						APPLICABILITY	SCORE
	Factor	Poor	Weak	Fair	Marginal	Qualified	Outstanding		
No.	*Description*	10	8	6	4	2	0	(A)	RxA
* 1	Does the Quality Data Program effectively support the six control elements of the Quality System?								
2	Does the Quality Data Program effectively support the supplementary measurement elements?								
3	Are forms properly engineered and managed for greatest economy and utility?								
4	Is optimum use made of computers for automatic analysis and feedback of quality data?								
5	Do all appraisal activities contribute properly to the Quality Data Program?								
* 6	Are sufficient quality data effectively transformed into information for all concerned, highlighting exceptions?								
* 7	Is there a comprehensive plan for software development, providing for efficient appraisal at each stage—with all data analyzed and used for quality improvement and reliability prediction?								
* 8	Are statistical techniques used to control the programming process and drive toward perfection?								
* 9	Is the software customer fully and rapidly supported during the life of the product?								
								100	
								Subsystem Rating	

* See page 280.

EXHIBIT 7 Evaluation Work Sheet: Special Studies

ORGANIZATION:

DATE:

SUBSYSTEM: Special Studies

Factor No.	Factor Description	Poor 10	Weak 8	Fair 6	Marginal 4	Qualified 2	Outstanding 0	APPLICABILITY (A)	SCORE RxA
* 1	Is Classification of Characteristics properly applied and communicated?								
2	Are disposition decisions properly made according to the Classification of Nonconformities program?								
* 3	Are environmental studies made and the results acted upon constructively?								
4	Are Machine and Process Capability studies made when needed and the results used properly?								
5	Are sufficient projects conducted to improve the general quality levels of products, processes, and service?								
* 6	Are all pertinent methods of statistical problem-solving and quality control properly, effectively, and sufficiently used?								
7	Are participative problem-solving techniques used regularly throughout the organization?								
* 8	Are all products certified and demonstrated to be reliable and safe when released for production?								
								100	
	Subsystem Rating								

Factor Rating (R)

* See page 280.

EXHIBIT 8 Evaluation Work Sheet: Quality Measurement and Control Equipment

ORGANIZATION:

DATE:

SUBSYSTEM: Quality Measurement and Control Equipment

	Factor	*Factor Rating (R)*						APPLICABILITY	SCORE
		Poor	Weak	Fair	Marginal	Qualified	Outstanding		
No.	*Description*	10	8	6	4	2	0	(A)	RxA
* 1	Are calibration and maintenance facilities adequate?								
* 2	Are calibration and maintenance programs fully satisfactory and effective?								
3	Are calibration and maintenance personnel fully qualified and in sufficient quantity?								
* 4	Is quality measurement and control equipment up to date, effective, and sufficiently integrated with production equipment?								
5	Is all quality measurement and control equipment properly documented?								
6	Are all tools and fixtures that are used as media of inspection fully qualified and identified?								
								100	
							Subsystem Rating		

* See page 280.

EXHIBIT 9 Evaluation Work Sheet: Human Resource Involvement

ORGANIZATION:

DATE:

SUBSYSTEM: Human Resource Involvement

No.	*Factor Description*	*Factor Rating (R)* Poor 10	Weak 8	Fair 6	Marginal 4	Qualified 2	Outstanding 0	APPLICABILITY (A)	SCORE RxA
1	Are all personnel fully familiar with their Quality System roles?								
* 2	Can all personnel who contact customers properly reflect Quality programs and status to them?								
* 3	Can all personnel who contact suppliers properly reflect Quality programs and status to them?								
4	Are sufficient personnel participating in professional Quality society and growth programs?								
* 5	Are training programs regularly conducted in all necessary Quality System subjects?								
6	Are the results of training properly evaluated and indicated program changes made?								
7	Is there sufficient rotation of personnel?								
* 8	Are proper performance standards participatively developed and regularly applied for all Quality System personnel?								
9	Are Quality System programs and problems regularly publicized to all personnel?								
								100	
	Subsystem Rating								

* See page 280.

EXHIBIT 10 Evaluation Work Sheet: Customer Contact

ORGANIZATION: DATE: SUBSYSTEM: Customer Contact		*Factor Rating (R)*						APPLICABILITY	SCORE
	Factor	Poor	Weak	Fair	Marginal	Qualified	Outstanding		
No.	*Description*	10	8	6	4	2	0	(A)	RxA
* 1	Are products properly labeled for user safety?								
* 2	Are installation and service difficulty, product failure, and repeat service information properly and rapidly transmitted to the factory and acted upon?								
* 3	Does the measurement of customer satisfaction reflect a proper level and trend?								
4	Are enough field failures fully analyzed and the results acted upon by the factory?								
5	Are field failure and other customer-related costs properly charged within the accounting system?								
6	Do field data sufficiently and accurately provide reliability information to product design?								
7	Are renewal parts of satisfactory quality when used in customer equipment?								
8	Are all field documents fully satisfactory at point-of-use?								
* 9	Are field quality activities properly audited and the results acted upon?								
								100	
							Subsystem Rating		

* See page 280.

EXHIBIT 11 Evaluation Work Sheet: Subsystem Summary

<table>
<tr><td rowspan="4">ORGANIZATION:

DATE:

Subsystems</td><td colspan="6">Subsystems Ratings (R)</td><td rowspan="4">RELATIVE IMPORTANCE (I)</td><td rowspan="4">(RxI) SCORE</td></tr>
<tr><td>Poor</td><td>Weak</td><td>Fair</td><td>Marginal</td><td>Qualified</td><td>Outstanding</td></tr>
<tr><td>1000</td><td>900</td><td>700</td><td>500</td><td>300</td><td>100</td></tr>
<tr><td>901</td><td>701</td><td>501</td><td>301</td><td>101</td><td>0</td></tr>
<tr><td>1 Quality System Management
2 Product Development Control
3 Purchased Material Control
4 Process Development and Operation Control
5 Quality Data Programs
6 Special Studies
7 Quality Measurement and Control Equipment
8 Human Resource Involvement
9 Customer Contact</td><td></td><td></td><td></td><td></td><td></td><td></td><td></td><td></td></tr>
<tr><td colspan="7">TOTAL</td><td>9.0</td><td></td></tr>
<tr><td colspan="7">System Rating</td><td colspan="2"></td></tr>
<tr><td colspan="7">Previous System Rating</td><td colspan="2"></td></tr>
</table>

System Rating:

9,000–5,001	Completely unsatisfactory
5,000–2,701	Marginally satisfactory
2,700– 901	Qualified
900– 0	Outstanding

Appendix A

Sample Procedure

The procedure selected to illustrate the ideas associated with this key type of document is one that tells how to produce a standard operating procedure. It incorporates ideas contained in the book *The New Playscript Procedure: Management Tool for Action,* by Leslie H. Matthies, published in 1977 by Office Publications, Stamford, Connecticut. Playscripting (presenting the material as a play is commonly written) enhances the clarity, completeness, and ease of understanding and use of the necessary directive information, particularly when it is accompanied by a well-organized flow chart.

Matthies has described (pp. 70–71) the basis for the Playscript concept as follows:

> Our organizations literally throb with life and action. Most of this human activity falls into specific, identifiable system channels. The actions that people take get results.
>
> Machines don't get results. Computers don't get results. Procedures and forms don't get results. PEOPLE DO.
>
> People make the organization hum. And dramas are being played every day and every minute in our organizations.
>
> Consider the group of actors and actresses in a movie or on television, or in the theater. They work together. Together they put on a performance and get a result. Some actors play major roles, some minor. Whether a woman plays the heroine or a maid, if the play itself is to get a result (be a success) each actress must do her bit well. Well-executed roles, plus a well-thought-out plot (plan for action) make a successful play.
>
> Each actor knows specifically what lines he is to speak and what action he is to perform. He even knows where on the stage he is to stand while he does his part. He knows how his part of the action relates to the parts of other actors and actresses. He knows when he is to come "on stage" and when he has done his job and when he is to exit.
>
> The sequence in which each actor speaks or performs some action is written in a script . . . a *Playscript.* A script is really just a procedure for putting on a play.

A procedure develops Quality Policy decisions into a statement of "who (organization or person) does what, when." This provides the basis for consistent, compatible handling of routine tasks and exceptions throughout an organization, irrespective of how far-flung it may be. The tool for ensuring adherence to such requirements and for updating of procedures for improved effectiveness and relevance in changing circumstances is 1.14 Audit of Procedures, Processes, and Products.

With a little effort, almost anyone can improve the effectiveness of procedure writing by using Playscript. The following procedure is an example of the Playscript approach and is one that has proven to be quite effective in getting people to write useful (and used) procedures. The first three pages of the example are introductory, providing a view of the structure and extent of the procedure that follows. The fourth page is a flow chart that, properly prepared, can be used by those not directly involved in the action to understand the sequence of actions. Beginning on the fifth page (page 4 of 13), the step-by-step Playscript process is shown.

One feature of the first page is the identification of a "Proprietor" for the procedure. The actual Proprietor is usually the top manager of the function designated on the procedure. It is his responsibility, delegated by the CEO of the organization of whose Quality System the procedure is a part, to ensure that the procedure is followed, modified as necessary, and has the intended effect on a continuing basis. The latter assignment is irrespective of those functions given action responsibilities within the Statement of Work of the procedure. As would be expected, then, the Proprietor function is usually the organization that has the greatest amount of important work to do as specified in the procedure.

Date: ____

STANDARD OPERATING PROCEDURE PREPARATION

Contents

PROPRIETOR: ____________________

PREPARED BY: ____________________ Date: ________

APPROVED BY: ____________________ Date: ________

REVISION: ________ APPROVAL: ____________________ Date: ________

STANDARD OPERATING PROCEDURE PREPARATION

XX–10 PURPOSE

To establish a uniform procedure for the writing of Standard Operating Procedures (SOPs).

XX–20 SCOPE

This document applies to all Standard Operating Procedures within the XYZ Corporation.

XX–30 GENERAL INFORMATION

An SOP may be initiated for any ongoing or intermittent activity involving more than one department and must deal with a central activity of the departments involved. The General Information section should list applicable documents, if any, as its first entry.

Once a primary objective has been agreed upon, any person may write an SOP. An early involvement of the Methods & Procedures Department is encouraged so as to obtain a number for the new SOP.

Under no circumstances is an SOP to be published without the review and approval of the Methods & Procedures Department.

All intradepartmental (within a department) procedures and instructions are to be written in accordance with the SOPs.

Every SOP must be approved by the manager of the department originating the procedure and the manager of Methods and Procedures. For procedures involving many departments, the required approval signatures will be determined prior to drafting.

When specifying actions that are to be performed, use playscript for the entire SOP. Playscript is always used, except when tables or number systems are the primary subject of the SOP.

A Flow Chart, outlining the general pattern of events in the procedure, is to be incorporated wherever playscript is used.

Playscript procedures are to be divided into the following sections: Purpose, Scope, General Information, Flow Chart, Work Statement, and Special Circumstances. See Exhibit D.

A procedure must be all-encompassing, if at all possible. Exceptions to this rule, which cannot be logically included in either the work statement or the flow chart, are to be covered in a "Special Circumstances" section following the work statement.

STANDARD OPERATING PROCEDURE PREPARATION

XX–30 GENERAL INFORMATION

The work statement is generally divided into several sections to cover all aspects of the procedure. Each section may be subdivided into subsections, if the writer determines that information which is not part of the main flow (but is still pertinent to the procedure) needs to be included.

In many cases, two situations that follow virtually the same flow are found in a procedure. Rather than write the entire procedure twice, use reference steps. For example, if two situations follow the same flow except for the initial step, write the full procedure for the first situation. For the second, write the initial step and refer the reader back to the first situation—for example, "see Steps 10.0 through 50.0."

Refer to Exhibit C, Glossary of Terms, for a better understanding of terms used in the procedure.

Date: ____

Revision Date: ____

Page 3 of 13

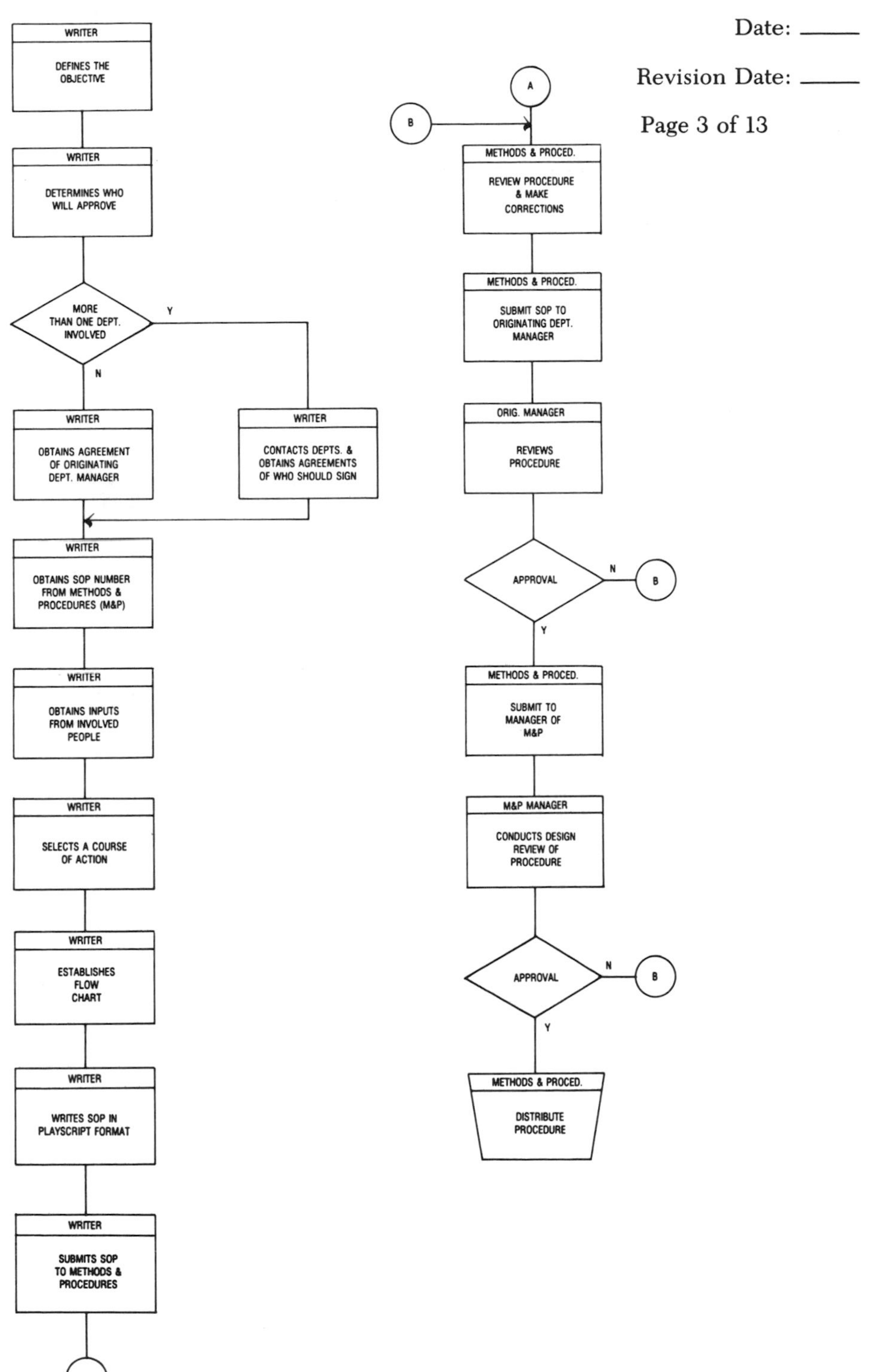

STANDARD OPERATING PROCEDURE PREPARATION

XX–40 Procedure Research

Responsibility		*Action*
Writer/Procedure Analyst	5.0	Defines the objective of the procedure and obtains the agreement of the originating department manager.
	10.0	Determines who will approve the procedure, if more than one department is involved.
	15.0	Contacts the Methods & Procedures Department to obtain an SOP number.
	20.0	Contacts as many involved people as possible for verbal or written inputs regarding the best course of action or what currently is being done.
	25.0	Determines whether there is conflict and coordinates a workable solution for all involved parties.
	30.0	Selects a course of action.

XX–50 Playscript Format

	35.0	Establishes a flow chart to help determine the sequence of events. See SOP AA–00, Flow Chart Standardization for SOPs.
	35.05	Uses the flow chart as a check list to determine if information is complete.
	40.0	Tells the reader how to proceed in action-oriented terms.
	45.0	Identifies the people or the organizations who perform the action. See Exhibit A.

STANDARD OPERATING PROCEDURE PREPARATION

XX–50 Playscript Format (cont.)

Responsibility		*Action*
Writer/Procedure Analyst	50.0	Writes the procedure in a logical sequence, numbering each step. *Note:* A suggested numbering system for SOPs is the numbering of each step by fives. This permits internal additions to be made without completely renumbering the ensuing action steps.
	55.0	Commits a specific person or organization to a specific responsibility.
	60.0	Begins each sentence with an action verb, unless an initial qualifier is essential for understanding (e.g., "If material is nonconforming, initiates MRB action." Without the qualifier before the action verb, confusion may result.) See Exhibit B. *Note:* For exceptions, see note to step 90.05.
	65.0	Uses present tense and lists descriptions in the form of a direction.
	70.0	Writes the procedure in as clear and concise a manner as possible. Uses simple words that leave no room for misinterpretation. Constructs the sentences in a straightforward manner, keeping the sentences short.

STANDARD OPERATING PROCEDURE PREPARATION

XX–50 Playscript Format (cont.)

Responsibility		*Action*
Writer/Procedure Analyst	75.0	Incorporates the "who does what when" format. See Exhibit A.
	80.0	Identifies the main channel of flow. Does not make a detailed list of intradepartment or job-description type functions.
	85.0	Makes liberal use of exhibits, placing them at the end of the procedure. Refers to exhibits in the written portion of the procedure.

XX–60 Playscript Variations

Responsibility		*Action*
Writer/Procedure Analyst	90.0	Shows any variation in the sequence by indenting the action paragraph. Keeps sequential substep or note indented until coming back to the main flow.
		90.05 Labels the substeps by decimal to distinguish them from the main action. *Note:* Substeps need not begin with an action word, but always use action words when an action is involved.

STANDARD OPERATING PROCEDURE PREPARATION

XX–60 Playscript Variations (cont.)

Responsibility		*Action*
Writer/Procedure Analyst	95.0	Refers to other procedures when substeps require numerous steps. See SOP AA–00, Flow Chart Standardization for SOPs.
	100.0	Incorporates notes to clarify certain steps, as needed.

XX–70 Procedure Approval

Responsibility		*Action*
Writer/Procedure Analyst	105.0	Submits procedure to Methods & Procedures Department for review.
Methods & Procedures	110.0	Arranges for and conducts design review meeting including representatives of all departments involved in the procedure.
	115.0	Obtains signature of person authorized to sign along with the signature of the Methods and Procedures Manager.
	120.0	Distributes procedure per SOP BB–00, Survey and Distribution of Standard Operating Procedures.

STANDARD OPERATING PROCEDURE PREPARATION

XX–80 Procedure Revisions

Responsibility		*Action*
Originating Department	125.0	Contacts the Methods & Procedures Department whenever a procedure becomes obsolete or outdated for cancellation or revision.
Methods & Procedures	130.0	Revises or cancels the procedure. *Note:* For procedures requiring major changes, the Methods & Procedures Department must have the changes reviewed and approved in the same manner as a new SOP. For minor changes, such as an assignment of a number or letter, the review and approval sequence is not necessary.

STANDARD OPERATING PROCEDURE PREPARATION

XX–90 Exhibits **Exhibit A**

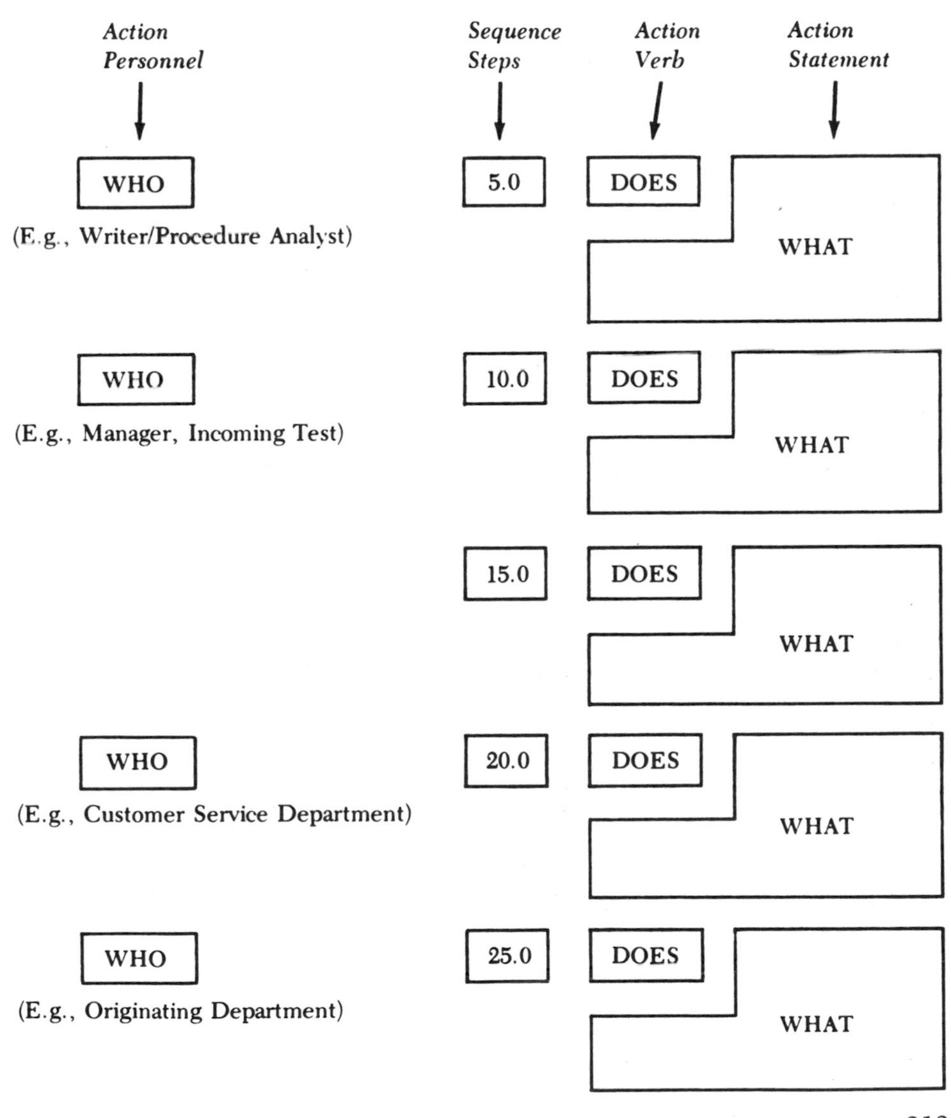

STANDARD OPERATING PROCEDURE PREPARATION

Exhibit B

PLAYSCRIPT ACTION WORDS

accumulate
advise
agree
aid
analyze
approve
arrange
ask
assemble
assign
assure
attach
authorize

begin
budget
build
buy

call
cancel
change
charge
check
claim
code
commit
compare
compile
compute
conduct
confer
conserve
construct
contact
control
coordinate
correct
count
create

define
delete
deliver
describe
design
destroy
detail
determine
develop
devise
distribute
document

edit
encourage
enforce
ensure
enter
establish
estimate
evaluate
expedite
explain

file
follow up
forecast
formulate
forward
further

gather
generate
get
give

handle
help
hire

include
incorporate
indicate
inform
initiate
insert
inspect
install
instruct
interface
interview
investigate
issue

keep
key

list
locate
log

mail
maintain
make
make available
mark
measure
move

note
notify

STANDARD OPERATING PROCEDURE PREPARATION

Exhibit B, concl.

PLAYSCRIPT ACTION WORDS

obtain
operate
originate
order

participate
pay
perform
pick up
place
plan
point out
post
predict
prepare
prescribe
present
prevent
print
process
procure
produce
project
proofread
protect
provide
prove
publish
pull
punch
purchase
push

read
recall
receive
reconcile
recommend
record
refer
reject
release
report
represent
request
require
resolve
restrict
retain
retrieve
return
review
revise
reword
route
run

sample
schedule
secure
see
select
sell
send
separate
serve
ship
show
sign
sort
stamp
state
stop
study
submit
supervise
supply
survey

tabulate
take
tape
tell
terminate
test
tie together
total
track
transcribe
transfer
type

update
use

verify

weigh
withdraw
work
write

STANDARD OPERATING PROCEDURE PREPARATION

Exhibit C

GLOSSARY OF TERMS

1. ***Action Words***—Active verbs, used in the form of a command, that begin each action paragraph, except as noted in XX–50, step 60.0.
2. ***Flow Chart***—A diagram that indicates the sequence of events and provides a basic outline of what is contained in the written portion of the procedure.
3. ***General Information***—Information that enhances the reader's understanding of the procedure but does not fit into the responsibility-action format. The general information section also describes ties with other programs and procedures.
4. ***Note***—Information that follows a step or substep and clarifies or enhances the reader's understanding of that particular step.
5. ***Playscript***—A method of writing consisting of two columns, one for responsibility and one for action. The advantage of playscript is its clarity; all parties involved have definite duties, presented in sequential order.
6. ***Purpose***—An explicit explanation of what the procedure is designed to accomplish.
7. ***Responsibility-Action Format***—The playscript method consists of two columns. The responsibility column designates the specific person(s) or organization(s) who performs the action. The action column defines the duties that the designated person(s) or organization(s) is to perform.
8. ***Scope***—Indicates what organizations, documentation, and/or activities are involved in or affected by the procedure.
9. ***Substeps***—Steps within the procedure that cover alternatives, extensions, or slight deviations from the main sequence of events.
10. ***Who Does What When Format***—The basic writing format for playscript. Eliminates all description and defines who performs an action and when that action is performed.
11. ***Work Statement***—Identifies people or organizations performing actions and provides a detailed explanation of what the actions are and of what is contained in the flow chart.

STANDARD OPERATING PROCEDURE PREPARATION

Exhibit D

PROCEDURE CONTENTS

Appendix B

Sample Work Instruction

The following page reflects the key features of an Inspection Instruction, which is one kind of Work Instruction. The Instruction itself is fully identified and authorized, including changes. The item being worked on is identified and illustrated, with the characteristics to be inspected the only ones shown. The application point of the Instruction is clearly stated, and all of the information the inspector needs to obtain the proper inspection equipment or laboratory help and then to perform the inspection is provided.

Rather than issue this information on a form for each inspection situation, some companies provide it through a computer display or printout. The display technique ensures immediate response to any changes.

To eliminate the need for an Inspection Report form in addition to the Inspection Instruction, additional columns are sometimes provided in the lower portion of the form. These columns often are as follows: Inspection Results, Inspector Identification, Date Inspected, and Lot Disposition. Obviously, less space would be available for some of the entries, but the full information associated with each lot would be available as part of the supplier part history.

INSPECTION INSTRUCTION NO. ____________ ☐ Incoming ☐ In-Process ☐ Final Page ____ of ____

Part No. ____________

Part Name ____________

____

Sampling Reference ____________

____

Prepared by ____________

Part Sketch

Supplier Name ____________ No. ____________

Laboratory Required ____________

Equipment No. ____________ Line No. ____________

Revisions			

Approved by ____________ Date ____________

Operation No.	*Char. No.*	*C/C**	*Characteristic Description*	*Limits*	*Sample Size*	*Inspection Equipment/Method*	*Equipment Numbers*

* Characteristic Classification

C = Critical
M = Major

N = Minor
I = Incidental

Index